The Nature of
the Meadowlands

Dedication

This book is dedicated to all the men and women whose vision and determination have helped the Meadowlands go from Jersey joke to New Jersey jewel.

Type set in Garamond Pro

ISBN: 978-0-7643-4186-1
Printed in China

All photography in this book is by Jim Wright unless otherwise noted

Cover and title page photos by Marco Van Brabant

Schiffer Books are available at special discounts for bulk purchases for sales promotions or premiums. Special editions, including personalized covers, corporate imprints, and excerpts can be created in large quantities for special needs. For more information contact the publisher:

Published by Schiffer Publishing Ltd.
4880 Lower Valley Road
Atglen, PA 19310
Phone: (610) 593-1777; Fax: (610) 593-2002
E-mail: Info@schifferbooks.com

For the largest selection of fine reference books on this and related subjects, please visit our website at **www.schifferbooks.com.**
We are always looking for people to write books on new and related subjects. If you have an idea for a book, please contact us at proposals@schifferbooks.com.

This book may be purchased from the publisher.
Please try your bookstore first.
You may write for a free catalog.

In Europe, Schiffer books are distributed by
Bushwood Books
6 Marksbury Ave.
Kew Gardens
Surrey TW9 4JF England
Phone: 44 (0) 20 8392 8585; Fax: 44 (0) 20 8392 9876
E-mail: info@bushwoodbooks.co.uk
Website: www.bushwoodbooks.co.uk

The Nature of
the Meadowlands

Jim Wright New Jersey Meadowlands Commission

With a Foreword by Governor Thomas H. Kean

Schiffer Publishing Ltd

4880 Lower Valley Road • Atglen, PA 19310

A Meadowlands sunrise, with the Manhattan skyline in the distance. *Photo by Ron Shields.*

Contents

An adult Bald Eagle perches in the Kearny Marsh, with Phragmites and Marsh Hibiscuses in the background. *Photo by Ron Shields.*

Foreword

Thomas H. Kean
Governor of New Jersey, 1982-1990

It is common to have a vision of something better, but most visions never come to pass. This is the story of one that did.

When I was growing up, the Meadowlands was a New Jersey joke. New York had dumped hazardous materials there for decades. In those days, very little wildlife could survive in that polluted environment. From the turnpike, you could see smoke billowing up from dozens of underground fires. Rumor had it that the Meadowlands had become the last resting place of Jimmy Hoffa.

One man saw the Meadowlands differently. Fairleigh Dickinson, Jr., was a businessman and a philanthropist who had the vision. His wife told me he ran for the New Jersey Legislature for only one reason: to see the Meadowlands restored.

He envisioned new jobs, the end of unregulated dumping, and the restoration of the marshes. To do this, he proposed local towns give up home rule to work together for the common good. At that time, his proposal – covering 30 square miles – was considered radical. But he never gave up.

Dickinson enlisted young legislators like Dick DeKorte of Bergen and Herb Rinaldi of Essex. Together, they created the Hackensack Meadowlands Development Commission with a mandate to restore the land in Bergen and Hudson counties.

The day the act passed, the land was so polluted that even environmental scientists declared it a dead zone. Very few thought it could be restored.

Now, 40 years later, everything is different.

In *The Nature of the Meadowlands*, Jim Wright chronicles this amazing story. In this once forsaken area are nesting Bald Eagles, Peregrine Falcons, and Ospreys. The fish are back in the Hackensack River. And thousands of schoolchildren every year visit a nature preserve in sight of the Manhattan skyline.

Wright's book shows once again nature's restorative power. It is a story of renewal that should be shared with everyone. Dreams can come true. With vision and hard work, almost anything is possible.

An aerial view of the Hackensack River, with Exit 16W of the New Jersey Turnpike and the Berry's Creek Canal in the foreground, and the eastbound Route 3 bridge on the left. *With aerial support from LightHawk.*

Introduction
A River Reborn

A river does not just happen; it has a beginning and an end. Its story is written in rich earth, in ice, and in water-carved stone, and its story as the lifeblood of the land is filled with color, music, and thunder.

— *Andy Russell,* The Life of a River

It's another hot June day in the Meadowlands of northern New Jersey, and the naturalists are glad to be aboard their 21-foot-long work skiff on the Hackensack River, far below the madding car and truck traffic that streams across the Route 3 bridges.

The jammed highways, the jumble of railroad tracks, the vast sports complex, the warehouses, and housing developments: these are what come to mind when most people think of the area. It is, after all, part of the most densely populated region in the nation's most densely populated state.

The New Jersey Meadowlands Commission's (NJMC) naturalists, who spend their summer days on the river and in the nearby marshes, know better. They know, first-hand, that another Meadowlands awaits – a place of daily surprises and small miracles.

On the river's banks, egrets and herons catch fish with their javelin beaks, and tiny Fiddler Crabs scuttle into the water as the naturalists' skiff zooms past. Above the long strands of Spartina grass in the nearby Saw Mill Creek Wildlife Management Area, a Northern Harrier wheels and scouts for prey while a distant airliner descends toward Newark Liberty Airport.

The day on the river is turning out to be a productive one. Already, the naturalists have cast a 50-foot-long net as part of a continuing fish study – and snared Blue-clawed Crabs and Diamondback Terrapins along with the Striped Bass and White Perch.

They train their binoculars on a small radio tower across the river from Laurel Hill County Park and confirm the reports of a pair of Ospreys tending to a ramshackle nest atop the structure, less than 50 yards from the rumble of trucks on the eastern spur of the New Jersey Turnpike.

The naturalists have even rescued a Double-crested Cormorant that they had initially given up for dead. The young web-footed bird somehow slipped from its perch on an old tide gate and wedged its neck in a gap between the weathered planks.

When they first spied the cormorant, it looked as stiff as the planks. But when they pulled alongside the tide gate and the summer intern gently extracted the bird, it suddenly sprung back to life and plunged into the river. Everyone gasped and nervously waited to see if the cormorant would surface again. Several moments passed. Then, 20 yards away in a nearby channel, the orange-billed bird's head popped out of the water like a cork, and a spontaneous cheer erupted. They had been part of a minor miracle.

But now, as they head back upriver minutes later, perhaps the biggest natural wonder of all unfolds before their eyes.

Somehow, a young Double-crested Cormorant had wedged its head between the gaps in an old tide gate.

Meadowlands Commission naturalists pull in a
50-foot-long fishnet on the Hackensack River.

The river's bounty on that day included Diamondback Terrapins, Striped Bass, and White Perch.

For the first time in more than a half-century, Peregrine Falcons are nesting on the bridges over the Hackensack River.

In the distance, two Peregrines come into view.

As their boat approaches the eastbound Route 3 bridge, two of the naturalists raise their binoculars and anxiously scan the tops of the bridge supports for Peregrine Falcons. They have seen these amazing raptors perched on the bridge many times before, but this time is different.

Earlier in the spring, a pair of Peregrines nested on the westbound Route 3 bridge, and the naturalists got fleeting glimpses of a makeshift nest and one – maybe two – baby Peregrines hopping around the top of the concrete bridge support. But in the past week, the naturalists have seen no falcon chicks, or adults. They worry that, as in years past, the falcon nest has failed.

Then, in the distance, two falcons come into view. Naturalist Mike Newhouse shouts the news over the thrum of the 115-horsepower outboard motor: "The one on the right looks to be a young first-year bird."

The naturalists point the boat toward the bridge abutment, then drift past at a respectful distance. All eyes are trained on the dark-feathered raptor above.

Newhouse is right. It is a recently fledged falcon, and cause for celebration. For the first time that anyone can remember, Peregrines have successfully nested in the Meadowlands.

For the falcons and the region, the young bird marks another milestone in an altogether remarkable recovery.

Fifty years earlier, Peregrine Falcons – one of the fastest creatures on the planet, with flight speeds approaching 200 miles an hour when dive-bombing prey – were virtually extinct east of the Mississippi River. The misuse of the pesticide DDT had weakened the shells of the falcons' eggs to the point where they'd break when the parents incubated them.

The Hackensack River and the entire 30 square miles of the Meadowlands District have much in common with the Peregrine Falcon. Written off as lost causes, both have made remarkable comebacks.

A half-century ago, the river had become so riddled with sewage and industrial contaminants that it had been pronounced dead. The adjacent landscape was equally ravaged. Marshes had become unregulated landfills where underground fires burned for years. The area stank of smoke and garbage for miles around. And the region developed a reputation so nasty that it has been hard to shake to this day: a wasteland of warehouses, industrial pipelines, midnight dumping, and buried mobsters.

The Meadowlands landscape will never return to the ancient days when Lenape tribes lived along the Hackensack, but the river and its tidal marshes have made a remarkable recovery that flies in the face of every half-baked Jersey joke.

Today, the Hackensack River supports so much aquatic life that the ramifications have reverberated up the food chain, from the tiny benthic organisms that live on the river's bottom to the schools of Striped Bass that swim in the river itself.

The Double-crested Cormorants see the river as one huge brackish buffet. Ospreys, a threatened species in New Jersey, now nest on several man-made structures on the river's banks. Peregrine Falcons now make their homes not just on the Route 3 bridges, but on other bridges that span the Hackensack River from Secaucus to Little Ferry. And Bald Eagles, seen regularly year-round nowadays, nest nearby.

The Meadowlands' recovery is far from complete, but it has reached the point where the time has come to look beyond the old stereotypes and take stock of the strides that the Meadowlands has made in recent years.

Ospreys have made an amazing rebound, with at least four active nests along the Hackensack River at last count.

The Meadowlands provides thousands of acres of open space on the doorstep of Manhattan and Jersey City.

This view of the Hackensack River from Laurel Hill includes the Saw Mill Creek Wildlife Management Area on left and the Upper Hack Bridge, a traditional lift bridge completed in 1959. *Photo by Marco Van Brabant.*

Chapter One
Getting a Sense of Place

In the past 30 to 40 years, nobody was deeply concerned over the acquisition of public land. Now there is very little land left, and it seems imperative we decide immediately on more open space.
– N.J. State Assemblyman Richard W. DeKorte
1973 interview with *The Ridgewood News*

The 30.4 square-mile Meadowlands District is located in northern New Jersey, roughly five miles west of Manhattan as the Great Egret flies. The district encompasses parts of 14 towns in Bergen and Hudson Counties. The Hackensack River runs through it, providing its environmental lifeblood.

The Meadowlands thrives with wildlife year-round. It is home more than 25 species of mammals (including an occasional Harbor Seal), more than 50 species of fish, more than 400 plant species, more than two dozen kinds of butterflies, and – at last count – 78 species of bees, including two that until a few years ago had never been seen before in the United States.

But that's part of the lure of the Meadowlands. You never know what you might see.

Because of its unique location on the Atlantic Flyway and its large expanses of mudflats and marshes, the region attracts an array of waterfowl, wading birds, and upland species. Migrants feed and rest in the spring and fall. Nesting species and winter residents also share the amazing open-air bed and breakfast that is the Meadowlands.

All told, some 280 species of birds spend at least part of their year here. These include almost half of the 77 birds on New Jersey's list of birds that are endangered, threatened or of special concern.

These bees are just one of the 78 species of bees that researchers have found in the Meadowlands.

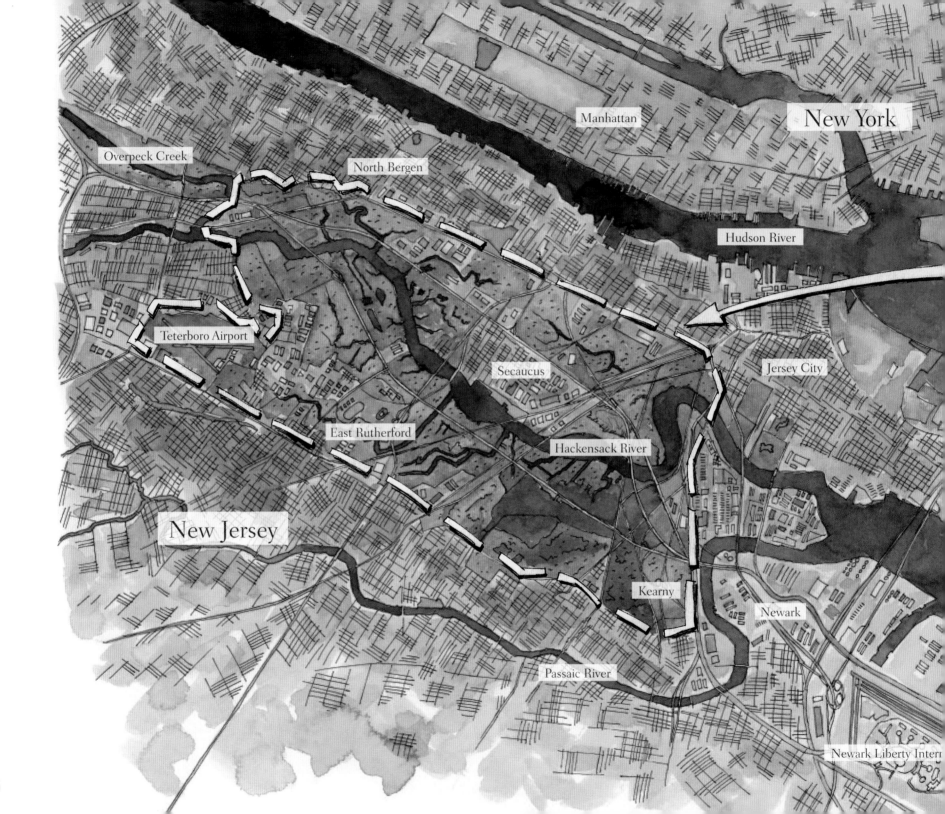

Overpeck Creek

North Bergen

Manhattan

Hudson River

Teterboro Airport

Secaucus

Jersey City

East Rutherford

Hackensack River

New Jersey

Kearny

Newark

Passaic River

Newark Liberty Intern

16

Brooklyn

Atlantic Ocean

Meadowlands

Upper New York Bay

Newark Bay

Staten Island

Airport

17

The Meadowlands is one of the few places in New Jersey that plays home to both Barn Owls and wild Ring-necked Pheasants. The Barn Owls live in old trash-baling facilities next to closed landfills and marshes, and hunt mice and other prey at night. The pheasants, which like to live near wetlands, thrive amid the Mugwort and high grasses on the closed landfills.

Each summer the district also hosts hundreds of the region's so-called Harbor Herons – Great Egrets and Snowy Egrets that breed on a deserted island in New York City's East River and then fly to the Meadowlands to feed and "loaf."

The region gets plenty of other unusual birds, from Snowy Owls, Rough-legged Hawks, and Northern Shrikes to Least Bitterns, Northern Wheatears, and Tri-colored Herons (see Chapter 5).

Wild Ring-necked Pheasants, rare in most of New Jersey, have found a home on at least two of the closed landfills in the district.

Great Egrets nest on a deserted island near Riker's Island in New York City in late spring and fly to the Meadowlands to feed.

The region's charms lie in the eye of the beholder. If you want to see industrial areas, warehouses, old landfills, highways, train tracks, and high-tension power lines, you will find them without much trouble.

But if you open your eyes to the natural wonders that also abound, then you will see rare birds and butterflies, vast tidal marshes, wonderful views of the Manhattan skyline, and storybook sunrises and sunsets.

Consider the perspective of a bird-watcher from South Jersey who visited the Meadowlands a while back and then announced on an e-mail list for birders: "I live out in the sticks where I can see the Pleiades just about any clear night and hear coyotes often in the woods behind my house. … I was exhausted when I got home [after visiting the Meadowlands] – not from birding, rather from my brain being totally overloaded with seeing so many cars, traffic lights, pavement and all that urban sprawl!!!"

He wrote about visiting ponds, marshes, and mudflats that "no sensible birders would visit – out-of-the-way areas along railroad tracks, behind industrial complexes." But he also wrote of all the wonderful species he had seen: "Black-bellied Plovers, Western Sandpiper, Gadwall, Green-winged Teal, American Coot, Swamp Sparrow, Short-eared Owl, and Barn Owl."

And that is the contradiction that is the Meadowlands. One naturalist's industrial wasteland is another naturalist's marsh. The Meadowlands is not located in a pristine wilderness. It is located on Manhattan's doorstep, in the fourth-largest metropolitan region in the entire world, and it must be judged in that light.

With more than 1,100 people per square mile, New Jersey is the most densely populated state (Rhode Island is a distant second, with just over 1,000 people per square mile, or 10 percent fewer). New Jersey's northeastern region, home to the Meadowlands, is the most densely populated part of this most densely populated state.

Some 8 million people live within a 10-mile radius of the Meadowlands, although most are aware of only the sports complex and outlet malls. To them, the region's marshes of the Meadowlands are worthless swampland – if they are on their radar screen at all.

Signs of humankind – from high-tension lines to airplanes – are never far from view. *Photo by Marco Van Brabant.*

Nothing could be farther from the truth. Researcher Joan Ehrenfeld of Rutgers University realized that urban wetlands had an importance overlooked by most. In an article for *Ecological Engineering*, she wrote that these marshes "often constitute the only 'natural' habitats that urban residents experience," and concluded: "These wetlands continue to provide important ecological services and may be of particular importance to both people and wildlife because of their remaining presence within concrete landscapes."

Look beyond the obvious human footprint, and you'll see all sorts of parks and trails. From north to south, here's a quick tour of the major natural areas, parks, and wetlands covered in this book.

The one thing that most of these open spaces have in common is they were once impassable for humans. In one way or another, the sites have been reclaimed, rehabilitated, or revitalized. Vast stands of the invasive common reed, a.k.a. Phragmites, have been replaced by native wetlands plants.

Marshes with lots of native Spartina grass provide great habitat for a variety of species, from fish to ducks.

Losen Slote Park, Little Ferry: This twenty-eight-acre natural area comprises six acres of meadow and a twenty-two-acre lowland forest – one of the few remaining historic parcels of woodlands in the entire district. The park is named for the Hackensack River tributary running through it. *Losen* is Dutch for "loose" (as in winding), and *Slote* is Dutch for creek.

In colonial days, the area was a semi-wilderness of marshes and wet woodlands, but it has become less wet as a result of the construction of dikes and tide gates at the turn of the twentieth century, to drain the wetlands and control mosquitoes.

The park's diverse environment supports many migratory songbirds, as well as a variety of small mammals and reptiles. On a summer's day, you might see an Eastern Box Turtle, an Indigo Bunting, or a flock of Cedar Waxwings. Nearby Mehrhof Pond was once a clay pit for a local brick factory. Now it is a major rafting site for wintering waterfowl, especially Ruddy Ducks, which congregate by the hundreds until the pond freezes solid.

The park, designed and constructed by the NJMC two decades ago, is owned and operated by the Borough of Little Ferry.

"This is really the only place we get to experience that last stand of lowland woods in the Meadowlands, with some amazing native plants like Summer Sweet and a nice stand of May Apple," says Don Torino of Bergen County Audubon Society. "That's why we get so many great warblers here – because of this habitat. Make no mistake about that."

Losen Slote Creek Park in Little Ferry features beautiful stands of birches and plenty of native plants.

Skeetkill Creek Marsh Park, Ridgefield: The sixteen-acre Skeetkill Creek Marsh, a small oasis in an otherwise light industrial section of town, was once heavily littered and filled mostly with invasive Phragmites.

The Meadowlands Commission restored the site in the late 1990s, constructing tidal channels and islands of native vegetation. In 2008, the commission turned it over to a non-profit group, the Meadowlands Conservation Trust, to maintain.

Peeps and other shorebirds flock to the marsh at low tide, and egrets and Great Blue Herons fly in to feed as the tide rises. Keep an eye out for warblers, Tree Swallows, an occasional American Kestrel and Northern Harrier, and – in autumn – Green-winged Teal.

A few hundred yards away on Railroad Avenue north of Skeetkill Creek Marsh is a colony of more than 30 Monk Parakeets, which have nested on a bridge above the railroad tracks for a decade. These tropical birds are the likely descendants of escaped or released pet birds, also called Quaker Parrots. The bright green birds have acclimated themselves to the Meadowlands in spite of the cold and snows of winter – and rank as the noisiest birds in the district.

A colony of Monk Parakeets has found a home on the Hendricks Causeway bridge abutments, building nests out of sticks.

Skeetkill Creek Marsh in Ridgefield is a small gem in a commercial part of town.

In the Richard P. Kane Natural Area, the invasive common reed Phragmites ise being replaced with native Spartina in an effort to revitalize the marsh.

This brand-new — and yet to be named — species of Leopard Frog was discovered in the Meadowlands and elsewhere in the New York metropolitan area. It lives in slightly brackish water and needs only small areas of adjoining uplands. *Photo courtesy Erik Kiviat/Hudsonia*

The Richard P. Kane Natural Area, Carlstadt and South Hackensack: This 597-acre marsh was for decades a vast stand of Phragmites and litter known as the Empire Tract, and slated to be the home of a huge shopping mall. In the wake of strong public opposition to the mall, the state bought the land and turned it over to the Meadowlands Conservation Trust.

In the past few years, much of the site has been restored to marshland, with new channels carved into the landscape to allow tidal flow. Hundreds of thousands of Spartina plants, a native wetland grass, replaced the Phragmites that once covered the land. Species ranging from Bald Eagles to a Harp Seal have been seen on the site.

An aerial view of Mill Creek Marsh (to the right of the eastern spur of the Turnpike) in Secaucus. *With aerial support from LightHawk.*

Mill Creek Marsh, Secaucus: Perhaps the most enduring impression of this 209-acre natural area is the view to the east just after you enter the site – from a shopping mall parking lot.

In the foreground, in one of the marsh's large tidal pools, are dozens of stumps of Atlantic White Cedars, the last vestiges of a primeval forest that once covered a third of the Meadowlands for hundreds of years.

In the distance stands the Manhattan skyline, a horizon's worth of skyscrapers, all of which were built after the last cedar was felled. That sort of juxtaposition – old and new, natural and man-made, side-by-side – is one of the distinctive defining features of the Meadowlands.

The marsh, covered by Phragmites since the early 1900s, was the proposed site for a development of 2,750 townhouses. In 1998, the NJMC restored the property, re-establishing tidal flows, creating open-water impoundments, and replacing the Phragmites with native plant species, in order to attract a diversity of aquatic life and birds.

The Secaucus Greenway includes a 1,500-foot elevated boardwalk that extends along the Hackensack River, from Secaucus High School to Mill Creek Point.

The rot-resistant cedar stumps provide perches for egrets and shorebirds in the two tidal pools. Green-winged Teal abound in the winter, along with Northern Pintails, Hooded Mergansers, and Ruddy Ducks. Summer is a time for sandpipers and yellowlegs and Snowy and Great Egrets.

The marsh meets the Hackensack River at Mill Creek Point, a three-acre public park built by the NJMC on the site of an old restaurant and marina. The park includes great river views, benches, and a launching ramp for canoes, kayaks, and other small watercraft. In the summer you'll see Ospreys and Forster's Terns. In the winter you'll see rafts of ducks, Bald Eagles, and an occasional Rough-legged Hawk.

To the park's south end is the 43-acre Secaucus High School Marsh, a former stand of Phragmites that has once again become a productive marsh. The marsh features a 1,500-foot-long elevated boardwalk that enables visitors to walk with the marsh on one side and Hackensack River on the other. The restored urban wetland attracts all sorts of birds – from rails to Saltmarsh Sparrows.

Mill Creek Marsh, with its Daylilies of early summer, is beautiful even at low tide.

River Barge Park, Carlstadt: One of the best ways to experience the Meadowlands is by boat. But when the river's water quality plummeted, many of the marinas that dotted the shoreline disappeared, to the point where the public no longer had any place to launch a boat on the western side of the river.

The opening of the NJMC's five-and-one-half-acre River Barge Park in 2012 changed all that. The marina, which includes the first dredging in the Hackensack River in more than twenty-five years, has a public boat ramp, ten-foot-wide docks and a special launching area for kayaks, canoes, and racing sculls.

The park also features an open-air environmental education pavilion (with a thatched roof made from Phragmites), a promenade overlooking the river, a parking area and a two boat-storage facilities. A pair of Ospreys has nested just to the north in recent summers, and the sandpipers resting on the docks are tempting targets for the local Peregrine Falcons.

River Barge Park's boat ramp provides the first public access to the Hackensack River from Bergen County in decades.

River Barge Park in Carlstadt, a riverfront park and marina, has buildings made with recycled materials – including 100-year-old timber that had been used for railroad barges.

Richard W. DeKorte Park, Lyndhurst: The Meadowlands Commission built this square-mile park, and its administrative offices, on the edge of the Kingsland Landfill in the 1970s, when Bergen County government wanted to extend the landfill all the way to the western spur of the New Jersey Turnpike – and fill in all the wetlands and mudflats along the way. By building its headquarters on the edge of the landfill, the commission sent a clear message: the era of using the region's marshes as dumping grounds was over.

The park includes several miles of trails, an environmental center, a butterfly garden, a low marsh, a LEED (Leadership in Energy and Environmental Design) platinum-certified science center complete with observatory, and two tidal impoundments that become vast mudflats twice a day.

Thanks in part to its tidal marshes, mudflats, uplands areas, and woods, DeKorte draws an amazing array of birds each season – from Northern Harriers and a dozen species of ducks in the winter to thousands of shorebirds and wading birds in summer and early fall. *BirdWatching Magazine* has named DeKorte Park and environs "among the top urban birding hotspots in the country."

One unique feature of the park is its efforts to make it accessible to students and others with all sorts of disabilities. The hard-packed paths, for example, are all at least six feet wide to enable wheelchairs to pass each other in both directions. The half-mile-long Marsh Discovery Trail, a seven-foot wide boardwalk, equipped part of the way with handrails, allows the disabled to experience wetlands up-close. Seating is located along paths throughout the park, so those with such mobility impairments can take a rest. A sensory garden with easy wheelchair access engages all the senses. Audio kiosks provide self-guided tours of DeKorte Park for the visually impaired.

"We decided to modify the park and make it accessible to students with disabilities for a simple reason: the kids were being left behind," says Michele Daly, director of disability education at the Meadowlands Environment Center. "So we started partnerships. We meet with the schools, find out what the issues are, and provide all the accommodations that the kids need."

The Shorebird Pool attracts Great Egrets and Snowy Egrets by the dozens in late summer.

DeKorte Park's Marshview Pavilion sits above the Shorebird Pool, a prime birding spot.

If you're lucky, you might see a rare Sora by the water's edge in the spring. *Photo by Greg Miller.*

Harrier Meadow, North Arlington: This 70-acre site, part of the Saw Mill Creek basin, was mostly Phragmites until the NJMC acquired it in 1996 and restored it two years later. The Meadow includes high marsh, meadows, three tidal impoundments, and a large tidal mudflat that attracts thousands of sandpipers and other shorebirds during migration – and plenty of ducks in winter. And yes, the wet meadows at Harrier Meadow are a favorite haunt of Northern Harriers, who nest nearby.

Harrier Meadow in autumn features Goldenrod and Groundsel at full throttle. The birding blinds provide views of three tidal impoundments.

Saw Mill Creek Wildlife Management Area: This 741-acre site is the largest expanse of tidal wetlands in the Meadowlands. The site had been diked and drained in the early twentieth century in a failed experiment in mosquito control. As a result, invasive Phragmites spread on the drained marsh soils and overran the fresh-water site. A nor'easter in 1950 destroyed the dikes, and tidal flow from the brackish Hackensack River was restored.

Native Spartina has slowly supplanted the Phragmites, and the Sawmill Creek marshes now have some of the highest biodiversity in the region. Critters that can be seen here in the warmer months include Diamondback Terrapins, Marsh Fiddler Crabs and Red-jointed Fiddler Crabs, herons, egrets, Black Skimmers, shorebirds, rails, and bitterns.

For the past few years, the Meadowlands Commission has been studying the terrapins in the marsh and other locations in the estuary. These medium-size turtles, considered an epicurean delight in the Victorian era, live in brackish estuaries and coastal rivers from Massachusetts to Texas. They are thought to have moved into the Saw Mill Creek marshes in the 1970s.

In recent years, terrapin numbers in these marshes have mushroomed. To monitor the turtle population in the Hackensack River estuary, each summer NJMC researchers briefly capture the terrapins, measure and age them, and then insert a small individually numbered RFID (radio frequency ID) tag transponder in each creature. In this fashion, researchers can determine when a terrapin is a "recapture" and make note of its location and condition.

A Montclair State University researcher has been taking blood samples of these turtles as part of a study to determine the genetic makeup and origins of turtles in the New York and New Jersey Harbor estuary.

(Opposite page)
The Saw Mill Creek Wildlife Management Area lies just to the west of the Hackensack River. This sunset shot was taken from Laurel Hill. *Photo by Marco Van Brabant.*

At low tide, Fiddler Crabs scurry around in the exposed mud of the Saw Mill Creek marshes.

A Meadowlands Commission study of Diamondback Terrapins in the Saw Mill Creek Wildlife Management Area captures the turtles long enough to record their measurements and age.

Researchers track the turtles via electronic tag numbers. The tag reader, encased in yellow plastic, is in the background. The turtles can live as long as 40 years or more.

Kearny Marsh, Kearny: This 310-acre impoundment, the only freshwater wetlands in the Meadowlands, was formed when the construction of embankments for railways and the New Jersey Turnpike cut off the area from tidal flow.

The marsh is located next to the once-notorious Keegan Landfill, which for decades leached all sort of pollutants, including heavy metals, into the marsh. The Meadowlands Commission has surrounded the landfill with a thirty-foot-deep clay wall and now pumps the leachate to a wastewater treatment plant.

The marsh itself is thriving once again, with Common Gallinules, Least Bitterns, Green Herons and raptors that range from Merlins to Bald Eagles. The marsh is best accessed by canoe and kayak.

A Red Slider (covered with Duckweed) and a Green Heron hang out in the Kearny Marsh, the only freshwater wetlands in the Meadowlands District. *Photo by Ron Shields.*

The view from the top of what's left of Laurel Hill includes ball fields in the park below, the Hackensack River, the Saw Mill Creek Wildlife Management Area, the western spur of the turnpike, the 1-E Landfill, and beyond.

Laurel Hill County Park, Secaucus: At first glance, the seventy-acre Laurel Hill County Park is a typical riverside park, featuring ball fields, playgrounds, a public boat ramp, and plenty of places where you can sit and enjoy the view of the Hackensack River and the vast Saw Mill Creek Wildlife Management Area beyond. An expanse of two-hundred-foot-high cliffs provides a stunning backdrop.

That rock formation is said to have inspired Prudential's "Rock of Gibraltar" logo, created by an advertising executive who saw it from a train in the 1890s. Much of that rock formation is long gone. Part of the hill was destroyed to make way for the New Jersey Turnpike in the 1950s, and much of the remainder was removed during quarrying operations in the 1960s and 1970s. For safety reasons, the cliffs are currently off-limits to the public.

For nearly a decade, the cliff has been home to nesting Common Ravens. When these huge black birds first nested, they were anything but common in the region. Nowadays, raven sightings are noteworthy but no longer rare. Other birds of note include Peregrine Falcons year-round and Great Cormorants, which perch on the nearby Hackensack River Swing Bridge in the winter. The past few summers, a pair of Ospreys has successfully nested just across the river by the eastern spur of the New Jersey Turnpike. The park contains a public boat ramp, a playground, lighted ball fields, and a riverfront walkway.

Common Ravens have nested successfully on the side of the cliffs at Laurel Hill for several years. *Photo by Marco Van Brabant.*

Two hundred fifty-foot-high Laurel Hill in Secaucus is the tallest natural feature in the Meadowlands. The eastern spur of the New Jersey Turnpike is to its right. *Photo by Marco Van Brabant.*

"The most amazing thing about Laurel Hill is that it's only a shell of what it was originally," says Secaucus Historian Dan McDonough. "There were 50 buildings here, including 11 hospitals. What remains of the hill is only a third or a quarter of what it once was. The hill used to come all the way out to close to the Hackensack River. Their water supply was a reservoir that was on top of this hill."

In that earlier life, Laurel Hill was known as Snake Hill, said to be a result of the large black snakes that were found there. As far back as the late 1700s, the site was the home to government buildings, beginning with a poor farm on the roughly 200-acre site. In the 1840s, Hudson County started locating other institutions there as well, including an insane asylum, orphanage, the hospitals, farm, three cemeteries, and a workhouse with a quarry where inmates crushed stones. Three churches were located on the hill as well.

By the 1930s, the institutions had begun falling into decline, and over the next few decades the buildings were shuttered and demolished – leaving that one huge smokestack and a few vestiges of the brick foundations of yore.

The county insane asylum on Laurel Hill stood for decades. *Courtesy of Dan McDonough – Secaucus Historian.*

Laurel Hill once housed several Hudson County institutions, three churches, and a pig farm. This view is from 1948; quarrying operations eventually reduced the hill's size by two-thirds. *Courtesy of the Hudson County Division of Planning.*

The Hudson County Alms House was built in 1863, when Laurel Hill was known as Snake Hill. *Courtesy of Dan McDonough – Secaucus Historian.*

HUDSON COUNTY PENITENTIARY.

The Hudson County Penitentiary featured a quarry where inmates used sledgehammers to crush rocks.
Courtesy of Dan McDonough – Secaucus Historian.

After the prison closed in the late 1950s, the Penitentiary Quarry was turned over to private quarrying operations with an eye to leveling the hill and turning it into a county park. The contractors allowed their mineral-collecting friends – a few at a time – on the site on weekends, and the rock hounds turned up some terrific finds, including Langite, Amethyst, Galena, and Serpierite.

The most incredible of the minerals was a microscopic piece of rock found by Nicholas Facciolla. In 1981, he took a sample of the unknown substance to the Paterson Museum for identification using a scanning electron microscope. In his pamphlet "Minerals of Laurel Hill," Facciolla described the mineral, less than millimeter in size, as "a hemispherical cluster of apple-green hexagonal crystals."

Facciolla's discovery was determined to be a new mineral, named Petersite for Thomas A. Peters, curator of minerals at the Paterson Museum, and Joseph Peters, curator of minerals at the American Museum of Natural History in New York City.

Langite is one of the microscopic minerals found at Laurel Hill. *Courtesy of Jack R. Troy.*

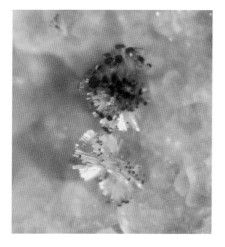

The microscopic mineral Petersite was discovered at Laurel Hill in 1981 by Nicholas Facciolla. He described it as "a hemispherical cluster of apple-green hexagonal crystals." *Photo by Brent Thorne.*

Rock collectors Nick Facciolla (left) and Jack Troy looked for minerals at Laurel Hill at a time when the old county buildings were being demolished. *Courtesy of Jack R. Troy.*

Chapter Two
In the Beginning

You have to know the past to understand the present.
– Carl Sagan, astronomer, *Cosmos*

First came fire, then ice, and eventually humankind. All three forces have left their indelible mark on the Meadowlands.

Volcanic activity first shaped the region millions of years ago. One of the few visible traces today is Secaucus's Laurel Hill, the mammoth chunk of igneous bedrock that juts out like a mammoth boulder plunked alongside the Hackensack River from the heavens above.

The Wisconsin Ice Sheet covered much of North America tens of thousands of years ago. The glacier, more than a half-mile thick in places, filled in the region between the Hudson River Palisades to the east and the Watchung Mountains to the west. The ice started to retreat twenty thousand years ago, eventually forming a vast freshwater lake called Glacial Lake Hackensack. The clay-bottom lake persisted for two-thousand years, until a breach developed and the lake drained.

Artist's depiction of the Meadowlands region (left center) at the time of the glacier's retreat, more than 12,000 years ago. *Illustration by John R. Quinn from his book,* Fields of Sun and Grass.

(Opposite page)
Hackensack Meadows at Sunset, by George Inness, 1859, Accession #S-22. *Collection of the New York Historical Society.*

Artist's depiction of the Meadowlands region roughly 10,000 years ago. Long ridge at right center is the Palisades, with Manhattan to its right. *Illustration by John R. Quinn from his book,* Fields of Sun and Grass.

Artist's depiction of mastodons along Glacial Lake Hackensack, roughly 11,000 years ago. *Illustration by John R. Quinn from his book,* Fields of Sun and Grass.

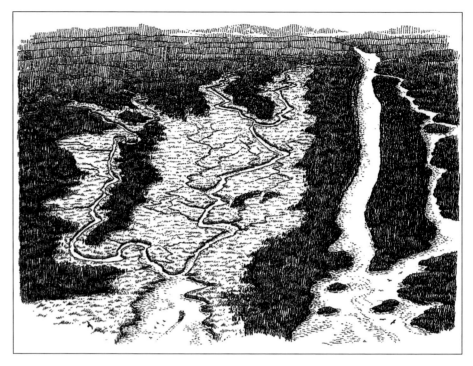

Artist's depiction of the Meadowlands region at the time of the Europeans' arrival in the early 17th century. *Illustration by John R. Quinn from his book,* Fields of Sun and Grass.

As John R. Quinn writes in his book *Fields of Sun and Grass*, "This was the age of the storied beasts of the prehistoric past: the moose-elk, a giant beaver weighing four hundred pounds, a twelve-foot-high sloth, peccary, caribou, the dire wolf, saber-toothed cats, the huge short-faced bear, woolly mammoths—and mastodons."

Over many millennia, the floor of the former Glacial Lake Hackensack collected water again, streams flowed southward, and marshes and forests took root. Sea level rose slowly, to the point where water from the Atlantic Ocean flowed inland. A coastal estuary developed, with the Hackensack River serving as its spine.

Native Americans that came to be identified as the Lenape arrived around 10,000 B.C., and found paradise along the banks of the river. They lived off the land and water, harvesting edible plants, hunting birds and mammals, and catching fish, turtles, clams, and oysters.

Atlantic White Cedars from the south made their way north circa 850 A.D. and soon became the dominant tree in the region. By several accounts, stands of Atlantic White Cedars – some up to 80 feet tall – grew in roughly a third of the Meadowlands at one time or another.

In *Discovering the Unknown Landscape: A History of America's Wetlands*, Ann Vileisis writes: "In swampy pockets from Florida north to New England, Atlantic White Cedars stood in austere cathedral stands. Requiring lots of light to germinate, these swamp trees generally grew in the aftermath of fires. Because native peoples routinely burned undergrowth and didn't suppress natural wildfires, white cedar forests flourished in pre-colonial America."

When Europeans arrived in the 1600s, those cedars and the rest of the region's bounty provided the natural resources to build settlements and eventually towns and cities. The settlers cleared the upland forests and converted them to farmland, and increasingly used the marshes to graze livestock and harvest salt hay.

Even so, the marshes of North Jersey were amazing places. As famed botanist John Torrey wrote nearly two-hundred years ago in *Catalog of Plants Growing Spontaneously within 30 Miles of the City of New York*, "Few places have afforded us more plants than in the vicinity of Hoboken and Weehawk, and the neighboring marshes."

According to *The Hackensack Meadowlands* by Kevin Wright, early Dutch settlers wrote of "mountain lions, bears, elk, deer, wolves, beavers, otters, fishers, catamounts, foxes, raccoons, minks, hares, muskrats 'about as large as cats,' martens, squirrels 'some of which can fly,' and woodchucks."

Settlers also began felling the rot-resistant Atlantic White Cedars, using the wood for shingles, plank roads, barrels, canoes, and other durable goods. In one instance, in 1791, thousands of acres of cedars in Kearny were burned to eliminate hiding places for pirates, who attacked ships in Newark Bay a few miles to the south.

Hundreds of rot-resistant Atlantic White Cedar stumps can be seen in Kearny, North Arlington, Lyndhurst, and Secaucus. This one is visible at low tide in DeKorte Park's Saw Mill Creek mudflats.

Atlantic White Cedar forests once covered as much as a third of the Meadowlands. This recent photo of a stand of White Cedars in the New Jersey Pinelands shows what parts of North Jersey likely looked like before the arrival of European settlers.

The Hackensack River Valley, so fertile and so near to New York City, became a breadbasket for the growing metropolis. Because the steep Palisades along the Hudson River prevented overland trade between North Jersey and the city, the Hackensack became a major regional thoroughfare, with commerce flowing back and forth with the tides.

The river was navigable all the way from its mouth in Newark Bay to Oradell, and sloops carried lumber and produce downriver and brought finished goods back up-river.

Valley of the Hackensack from the Estate of L. Becker, Esq, Union City, New Jersey, mid-19th to early 20th century, by Andrew Melrose. *Collection of the Newark Museum.*

Snake Hill on the Jersey Meadows, by Charles Parsons, 1871. *The Montclair Art Museum.*

The arrival of the railroads in the 1830s slowly changed the landscape further. No longer was commerce dependent on the tides or the season, and downtowns developed around railroad stations as well as along the river.

The marshes, at first considered an obstacle to the railroads, were conquered in part thanks to the abundance of white cedars. "The native growth of cedar on some parts of the marsh will furnish at once a cheap and durable timber," wrote civil engineer J.L. Sullivan in 1829 in his *Survey for a Rail-Road from Paterson to New York*.

By the late 1800s, railroads began to crisscross the Hackensack Meadows. When the automobile arrived at the turn of the century, it was no match for the marshes of the Meadowlands. As one letter writer complained in the automotive trade magazine *The Horseless Age* in 1903, the wonderful roads along the Watchung Ridge "could not be reached from the east without traveling over an imposing barrier of the most disgraceful and detestable thoroughfares for a distance of eight or nine miles through the Hackensack Meadows."

In part because of those rudimentary roads, trade along the river continued to flourish as well. One of the big manufacturing centers along the river was Little Ferry, where the local clay from old Lake Hackensack – when combined with sand, water, and coal dust – was found to be perfect for brick making.

A dozen brickworks were established, and the kiln-hardened bricks were shipped down the river. At the brickworks' peak in 1895, they produced one hundred million bricks a year, which became buildings in Newark, Paterson, New York City, and cities in New England. The last brickworks in Little Ferry endured until the 1950s.

As the human population expanded all along the East Coast, marshes were increasingly seen as worthless swamps, targeted for "reclamation" – a euphemism for the damming, diking, and filling them so they could be used for other purposes.

Judith S. Weis and Carol A. Butler write in *Salt Marshes*, "These projects were considered a useful way to decrease crowding in the cities and to increase the amount of fertile farmland, while at the same time ridding the world of mosquito-infested wetlands. Most of the city of Boston was built on salt marshes and tidal flats, as were much of Queens, significant parts of the Bronx and Brooklyn in New York City, and many towns in northern New Jersey near the Hackensack Meadowlands."

The Hackensack River began its slow descent, from serving as the region's Main Street to its back alley.

An 1832 timetable for the Paterson and Hudson River Railroad. The first cars were double-deckers, drawn by "fleet and gentle horses." *N.J. Meadowlands Commission archives.*

Little Ferry was one of the largest brick producers in the world in the late 1800s, exporting bricks down the Hackensack River to cities in the Northeast. *Courtesy of Little Ferry Historical Society.*

Mosquito Control Commissions built dikes and ditches more than a hundred years ago in a futile effort to rid the Meadowlands of saltmarsh mosquitoes. *N.J. Meadowlands Commission archives.*

Chapter Three
The Dark Ages

Unfortunately, our affluent society has also been an effluent society.
– Vice President Hubert H. Humphrey
Speech, October 11, 1966

In Europe, the Dark Ages lasted a half a millennium, beginning in 476 A.D.

In the Meadowlands, the Dark Ages lasted half a century, beginning shortly after World War I.

Although this swath of northern New Jersey had been settled steadily ever since the first Europeans took root, it changed drastically in the 1920s, when a decade-long prosperity fueled the popularity of the automobile and the growth of North Jersey suburbs.

As the region's human population expanded, so did the need for drinking water, and in 1921, the Hackensack Water Company began to build the Oradell Dam and dredge the river to its north. When the twenty-two-foot-high dam was completed two years later, the Hackensack became two rivers – with fresh water to the north of the dam and a brackish estuary of Newark Bay to the south. The Meadowlands would never be the same.

"At one time there were fresh-water tidal wetlands in the Meadowlands, and those are gone because of a greater intrusion of salinity," says Erik Kiviat, executive director of Hudsonia, a non-profit research institute for the environmental sciences. "That changes the entire flora and fauna, because a lot of species have upper salinity limits."

The first of the major Hudson River crossings, the Holland Tunnel, arrived in 1927 (followed by the George Washington Bridge in 1931 and the Lincoln Tunnel in 1937). No longer did automobiles need to travel by ferry, and New Yorkers looking to escape the big city had easy access to North Jersey's suburbs. The decade wasn't called the "Roaring Twenties" for nothing – the region's population expanded by 75 percent during that time.

But the Regional Plan of New York and Environs set its sights on a far more ambitious goal. An article in the October 1928 issue of *Popular Science* outlined the association's blueprint: "From a vast … waste of mosquito-infested swamp land, just across the Hackensack River from New York City, soon may rise a great city of industries and homes, larger than New York herself."

The plan included raising Hackensack Meadows ten feet by filling it with 200 million cubic feet of dirt, and straightening and dredging the Hackensack River for large ships. According to the plan, the city would accommodate 780,000 people, "about equal to that of Boston."

Manhattan Island from Jersey Meadows by William C. Palmer, 1934. *Smithsonian American Art Museum.*

43

The Great Depression – and reality – torpedoed any chance that plan ever had, but after World War II, the population exploded again. Soon, the boom towns of North Jersey had to address two increasing problems: their sewage and their garbage. For the towns along the Hackensack River, the answers to both were in their backyard. They could dump the trash into the nearest swamp and pipe the sewage into the river or its tributaries.

"The mentality before the Clean Water Act arrived in the 1970s was that swamps were worthless pieces of property where people dumped the things they didn't want to deal with – sewage and garbage," says Tom Marturano, the Meadowlands Commission's Director of Solid Waste and Natural Resources.

According to the U.S. Fish and Wildlife Service, the Meadowlands lost 14,000 acres of wetlands between 1889 and 1985 – mostly for development and landfills.

One of the region's most notorious dumps was the Keegan Landfill in Kearny, where just about anything and everything was dumped, including chromium by the drum-full.

Marturano says that when the Meadowlands Commission adopted the orphaned Keegan Landfill, it discovered an underground lake of oil. "In the old days, one of the ways that the landfill's operators controlled dust was to take waste oil from companies that generated it, dump it into a pit in Keegan, and other trucks would suck it up and spray the oil on the roads to keep the dust down."

Another major problem with these unregulated landfills was the toxic tea they generated, called leachate, which seeped into the ground and adjacent waterways after every heavy rain.

In her book *Garbage Land*, author Elizabeth Royte describes leachate as "a noxious stew of household toxics, such as battery acid, nail polish, pesticides, and paint, combined with liquid versions of rotting food, pet feces, medical waste, and diapers. … Analyses of leachate from Meadowlands landfills have turned up oil, grease, cyanide, arsenic, cadmium, chromium, copper, lead, nickel, silver, mercury, and zinc."

Stephen Quinn, senior project manager for the American Museum of Natural History, grew up in Ridgefield Park, just to the north of the Meadowlands District. He remembers how in the 1960s the monstrous landfills of Lyndhurst and North Arlington "always cast an ominous and dark spell on all of us who loved birding and nature. You knew the land was not long for this world. At that time there was no New Jersey Meadowlands Commission, and there were no decisions being made regarding logical development.

"When I hiked across the landfills as a kid there were hot streams of leachate flowing down the landfills and out into the surrounding creeks – they were just this god-awful-looking soup. In the wintertime I'd often find huge flocks of starlings that had bathed in the hot leachate and then fly to nearby Phragmites to dry off, and if it was really cold they would freeze to death."

Another horrendous problem for the landfills and surrounding Phragmites was fires.

"Back in the late 1950s, where the Sports Complex is, a landfill used to burn for years," says Don Smith, a retired NJMC naturalist who grew up in nearby Little Ferry. "You'd drive by there and there'd always be the smell of burning garbage. The dump was closed by then, but there was never any action by the county or anybody else to put the fire out. They just let it smolder."

In the 1940s and 1950s, fires were set on the landfills to incinerate the trash, but the fires continued for years after the practice was stopped. "More often than not you had all kinds of flammable junk being dumped there, you had spontaneous combustion, and you had the sun hitting a piece of broken glass like a magnifying glass and starting a fire," says Smith. "You had different sources for the initial ignition, but the consequences were the same. The fires burned deeper and deeper into the landfill."

The fires persisted because towns didn't have the money to pay for the big equipment needed to extinguish them.

In the old days, the garbage dumps attracted thousands of gulls.
N.J. Meadowlands Commission archives.

The problem finally came to a head in late 1973, when a combination of an inverted air mass, steam from the heated discharge water at the PSE&G plant, and a persistent fire from an abandoned landfill touched off a cataclysmic series of collisions on the New Jersey Turnpike in East Rutherford one October night. Although the smoke and fog reduced visibility to virtually zero, the stretch of roadway remained open for several hours – long enough for collision after collision to occur.

Ron Coldon of East Rutherford was one of four first responders on the scene. He recalls they could barely see a few feet in front of them as their ambulance crept down the roadway from Exit 16W: "As we went in, we kept hearing one deafening bang, then another and another – the vehicles kept crashing into each other. When we got there, dead bodies were everywhere. We pulled one out of the back of a garbage truck. We pulled one out of the swamps. One was lying dead on the highway. It was the most horrifying thing I ever saw... It was like the end of the world out there..."

The National Transportation Safety Board, which investigated the horrific crashes, issued a grim tally: "In all, 66 motor vehicles were involved; 9 persons were killed, and 39 others were injured."

The NTSB also issued a sweeping series of recommendations, including that federal and New Jersey state agencies "eliminate the possibility of fire and smoke from the old dumps within the Hackensack Meadowlands."

For much of the twentieth century, the Hackensack River was in increasingly sad shape as well. Seven sewage plants provided "primary treatment" – a euphemism for gravity. Sewers drained into holding tanks, where the heaviest effluent sank and the rest went into the river. Every time heavy rains overwhelmed these rudimentary facilities, the Hackensack River became an open sewer.

"That massive oil spill in the Gulf of Mexico in 2010 leaked 220 million gallons of crude oil over a three-month period," says Dr. Francisco Artigas, director of the Meadowlands Environmental Research Institute. "By comparison, the rudimentary sewage treatment plants along the Hackensack River dumped 115 million gallons of diluted sewage daily, for years."

(NY19)KEARNY,N.J., Oct. 24--AFTERMATH OF TURNPIKE TRAGEDY--Low level helicopter view shows wrecked vehicles piled against each other Wednesday morning following a major accident on the fog-shrouded New Jersey Turnpike at Kearny. At least nine persons were killed and more than 40 injured in three major pileups and scores of minor accidents on the turnpike, police said. Fog still shrouds the background.(AP Wirephoto)(see AP AAA wire story)(pr4ll48stf-DP)1973

A combination of fog and smoke from a fire on an old landfill contributed to a tragic series of crashes that involved 66 vehicles and killed nine people on the western spur of the New Jersey Turnpike in 1973. *Courtesy of Associated Press.*

Industry turned the Hackensack River and its creeks into sewers as well. In one instance just to the north of the district in Ridgefield Park, a paper factory extended over the river, and the floor drains emptied directly into the Hackensack.

On the other side of the river, manufacturers of disposable medical supplies, pharmaceutical products, and organic chemicals dumped so much waste into Berry's Creek, a tributary of the Hackensack, that it is still one of the worst mercury contamination sites in the world. Over one 45-year stretch, a mercury-processing plant discharged an estimated 270 tons of highly toxic mercury into a 2,000-foot-long section of the creek.

Other industries left a legacy of pollution as well – dioxin, asbestos, benzene, and PCBs that poisoned the Hackensack River for decades.

The river became so polluted that owners of boats docked on the river did not have to worry about barnacles. The Hackensack's dissolved oxygen levels were too low for barnacles to live.

Making matters even worse, as recently as 1995, PSE&G's Ridgefield Park power plant's cooling system dumped approximately 645 million gallons of heated water into the river every day, also wreaking havoc on dissolved oxygen levels and what few fish remained.

Reinforcing that unsavory image were the notorious pig farms of Secaucus.

Secaucus Historian Dan McDonough estimates that from the 1930s into the 1950s, his town had as many as 50 pig farms at one time, with the larger ones housing up to 10,000 pigs weighing up to 250 pounds.

Although some claim the area stunk not because of the farms but the local rendering plants for the pork fat, the fact remained that most people kept their car windows rolled up when they drove through Secaucus even on the hottest days.

The region got additional boosts of questionable publicity throughout 1950s, when Secaucus pig farmer and tavern owner Henry Krajewski ran for president as in independent -- with a piglet under his arm. His campaign platform included "No piggy deals in Washington."

Joe McKay, a folk singer, grew up on a 2.5-acre pig farm on Secaucus Road, not far from Krajewski's farm and tavern. McKay, who as a boy stapled "Krajewski for President" posters on local telephone poles, remembers his childhood on his father's 6,000-pig farm fondly.

"I thought it was the greatest thing ever," McKay says. "All our neighbors were pig farmers, so we all had a lot in common, and there were a lot of taverns with music all the time."

The dissolved oxygen levels in the Hackensack River estuary were once so low that barnacles couldn't survive. These days they abound – even on old Atlantic White Cedar stumps.

Secaucus pig farmer Henry Krajewski ran for president of the United States in 1952, 1956, and 1960, capitalizing on his occupation and underscoring his town's unpleasant image in the process. *Courtesy of Harvey Sullivan IV.*

McKay says his father Ben would buy 30-pound piglets at 10 cents a pound, fatten them up, and then sell the 200-pound-plus pigs at the same price for pound. To feed all those pigs, McKay's father sent his own garbage truck into Manhattan each night to pick up food waste and table scraps from the Waldorf Astoria and a multitude of restaurants.

The garbage was dumped on the concrete floor of the pig stables before dawn, and when the pigs came out to eat, their pens were hosed down. The pig manure was flushed onto a conveyor belt and deposited on the edge of the marsh.

Most of the pig farms went out of business by 1960, but the other problems – sewage, rampant chemical dumping, landfill fires, leaking leachate – persisted. By the late 1960s, the Hackensack River had hit its lowest ebb.

Pig farms abounded in Secaucus through the first half of the 20th century – including one operated by Ben McKay. *Courtesy of Joe McKay.*

Contrary to urban legends, the only signs of Jimmy Hoffa to be found so far in the Meadowlands are his son's old Teamster campaign stickers, still adorning a light pole or two.

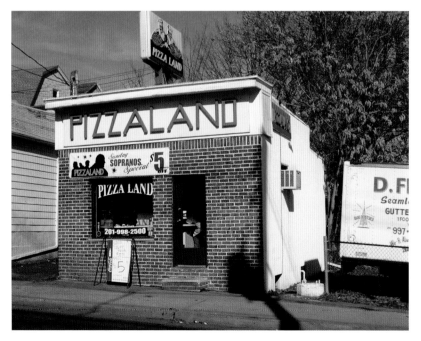

Pizzaland in North Arlington still gets fans of the old TV show "The Sopranos" stopping by to take photos of the building, featured in the opening credits.

A Few Words about the Mob

One of the most enduring tales about the Meadowlands is that it is a favorite mob burial ground. The most prominent example, of course, is Jimmy Hoffa, who was long-rumored to be buried in an end zone of the old Giants Stadium. (In fact, there is a "Hoffa 2001" campaign sticker on a traffic light less than two miles from the old stadium, although the Hoffa in question was the late Teamster leader's son.)

Organized crime has had long-time ties to the waste disposal industry and the thousands of acres of once-unregulated landfills.

"Towns were notorious for leasing out wetlands for fifty dollars an acre to private firms who'd fill them so that the town could get economic development out of them," says the NJMC's Tom Marturano. "Unfortunately, some of the people they leased it to were not the most upstanding businesspeople. When society has something that it's willing to pay anything not to deal with, that's when opportunities open up for people who are nefarious. The people who were willing to take care of society's crap made a lot of money. It didn't matter if it was sewage, garbage, hazardous waste – back in the early days there was no distinction."

The mob-Meadowlands connection has been reinforced by "The Sopranos," the classic HBO series about organized crime in North Jersey. The series' first season even featured an episode entitled – what else? – "The Meadowlands." Among the many references to the region: Pizzaland in North Arlington was featured in the show's opening credits, and one of Tony Soprano's favorite hangouts was the fictional Satriale's Pork Store in Kearny.

But the mob's ties to the region go back much farther. A century ago, when the Meadowlands was still mostly marshes, a 1910 *New York Times* headline proclaimed: "Hackensack Meadows a Hiding Place for Fugitives." According the lengthy article, the area was also a dumping ground for victims of a mob extortion scheme known as the Black Hand.

As a local police chief told *The Times*: "...in the Bergen County part of the Meadows a murdered Italian shows up every now and then. One of the last murder cases was discovered through a letter sent to Police Headquarters. It was written in Italian and said that a body would be found in a certain part of the meadows. … We got out and hunted for the body. It showed up all right. The man was murdered. Another Black Hand case, I reckon."

Chapter Four
Turning the Tide

In the 19ᵗʰ century, we devoted our best minds to exploring nature. In the 20ᵗʰ century, we devoted ourselves to controlling and harnessing it. In the 21ˢᵗ century, we must devote ourselves to restoring it.

— Stephen Ambrose, Historian

With the publication of *Silent Spring* in 1962, Rachel Carson's eye-opening book about the impact of pesticides on the natural world – from insects to humans – Americans began to view the natural world in a new light. Nature existed not just for the convenience of humans but to be shared with other creatures for the long-term well-being of all.

People started seeing wetlands as marshes, not swamps – places to be revered, not reviled. In that climate, in 1969 (a year before the first Earth Day), the New Jersey Meadowlands Commission was formed as the Hackensack Meadowlands Development Commission, with three seemingly contradictory mandates: "to protect the delicate balance of nature, to provide for orderly development, and to provide facilities for the disposal of solid waste."

By fighting pollution, it turned out, the commission could help accomplish all three.

Cleaning up the Hackensack River and its estuary was the logical place to start, because it was so contaminated. Don Smith, the retired NJMC naturalist, got his feet wet at the commission as a water-pollution inspector and saw the changes first-hand.

"I'd work with newly formed DEP's pollution branch," says Smith. "I'd pinpoint the source of the pollution, and the DEP would come up and take a sample to the state lab. If justified, we'd then build a case to remedy the situation. The Clean Water Act was passed in October 1972, and we already had 50 active cases for water pollution violations by then. Some were minor, but some were major – like foot-deep pools of arsenic behind metal-plating factories. Growing up, I used to wonder why there were no muskrats there, no ducks... I found out."

Slowly, thanks to money from the federal government, many of the sewage treatment plants along the river were upgraded. It took until 2010 to eliminate one of the last nasty sewage treatment plants, in North Bergen. The sewage now goes to Newark for treatment.

Combined-sewer overflows, another nasty source of pollution, are slowly being eliminated throughout the Hackensack River watershed. Rainwater runoff, raw sewage, and industrial wastewater all go into the same pipe to a sewage treatment plant. When the wastewater volume is too large – after heavy rains, for example – the treatment plant can't handle it all, and the overflow goes untreated into the nearest body of water.

In some marshes, native Spartina grass has replaced the invasive reed Phragmites, increasing biodiversity.

Ending the rampant dumping of trash, garbage, and other waste was another huge challenge. One of the commission's first decisions was a crucial one. Although conventional wisdom said that filling in wetlands was a way to get rid of garbage and create buildable land at the same time, the commission ruled that from then on, landfills would expand upward, and not into the marshes.

To that end, the commission rezoned thousands of acres, designating them for "wetlands preservation" so they could not be used for dumps or development. For the first time ever, state government recognized wetlands as a resource instead of a wasteland.

In 1981, the state enacted landfill regulations that are now standard procedures, such as prohibiting the dumping of often-toxic liquid waste. In light of the toxic inheritance contained in the orphan landfills it has adopted, the Meadowlands Commission has made it a practice to encase the landfills with clay six feet wide and sixty feet deep, and collects and treats the toxic runoff, known as leachate.

Another reason that the estuary is getting cleaner is the restoration of many of the remaining wetlands.

In the past two decades, the Commission (NJMC) and the Meadowlands Conservation Trust have replaced hundreds of acres of invasive Phragmites with more-desirable native vegetation.

"The goal is to help the wetlands better to perform their essential functions, including being productive, improving water quality, acting as a buffer during flooding, and fostering greater biodiversity," says Dr. Ross Feltes, the NJMC's Supervisor of Natural Resources Management.

The upgraded wetlands and the crackdowns on pollution have had a dramatic impact on the Hackensack River and wildlife – from the smallest benthic creatures that live in the mud on the river's bottom to the impressive Great Blue Herons who eat the estuary's fish.

A NJMC study a few years back found that the Hackensack River contains significantly fewer harmful metals than it did in the late 1980s. Although the mercury levels still exceed federal limits, there were substantial drops in the amounts of chromium, cadmium, nickel, copper, and lead.

The Meadowlands Environmental Research Institute (MERI), the Meadowlands Commission's research arm, has monitored the percentages of dissolved oxygen in the Hackensack River each season since 1993, and has found that the water quality has improved significantly. Dissolved oxygen levels – a key to sustaining aquatic life – have increased by 11 percent during that time.

A key early decision by the Meadowlands Commission was that landfills could expand upward instead of into the region's wetlands. *N.J. Meadowlands Commission archives.*

In the Richard P. Kane Natural Area in Carlstadt, the Meadowlands Conservation Trust has removed hundreds of acres of Phragmites and replaced it with plugs of native Spartina. That's MetLife Stadium on the right in the distance.

Efforts to improve the Richard P. Kane Natural Area have included reintroducing tidal flow from the Hackensack River and planting plugs of Spartina by hand.

The NJMC's data have also found a significant improvement in the benthic organisms – the mud-dwelling critters – in term of both numbers of species and numbers of organisms. A healthier benthic population has helped support the numbers and diversity of fish species in the river. Fish, in turn, feed birds, and studies by NJ Audubon and others show avian species are returning in large numbers to the region as well.

Some chronic problems persist. As of late 2011, several combined sewer overflows still emptied into the river and its tributaries just outside the district, including six in Hackensack and Ridgefield Park and a few others on Kearny.

In August 2011, Hurricane Irene inundated the Bergen County Utility Authority's plant in Little Ferry, spilling nearly 50 million gallons of sewage and runoff into the Hackensack River. The plant's capacity, 100 million gallons, was no match for the storm, and it took days to dissipate the pollution.

Nonetheless, the trend line has become clear. What started as a federal law in 1972 has finally become a public mind-set: Rivers should be recreational resources, not dumping grounds.

Although litter still enters the river and marshes – mostly from storm sewers and incoming tides – the estuary is cleaner most of the time. One major reason is the clean-up efforts of nonprofit groups and volunteers. State government recently issued water-quality regulations that require municipalities to install storm drains with better trash-catching grates, and that should ultimately reduce the amount of litter in the river even more.

Through guided nature walks and pontoon-boat cruises, more and more people have been experiencing first-hand a rejuvenated Hackensack River, and realizing why it matters – and that may be one of the most important factors in the river's recovery.

"The biggest change in the Meadowlands has been the attitude of the people who live near it," says Don Smith. "For too many years, the negative things that happened occurred because people didn't care. They didn't care because they didn't know. Now they finally understand."

As the Hackensack River has gotten cleaner, more and more fish – and more varieties of fish – have returned. This Oyster Toadfish was netted during a Meadowlands Commission fish survey. *Photo by Brett Bragin.*

Pontoon boat rides are part of a long-term effort to get the public to see the river and its wildlife up-close and take an interest in the estuary's future.

Chapter Five
Rare Birds and Odd Ducks

I never for a day gave up listening to the songs of our birds, or watching their peculiar habits …
– John James Audubon

The challenge in writing about the birds of the Meadowlands is where to begin – there are some 280 species and counting, with so many locations and so many types of habitat within a half-hour's drive.

Because of the Meadowlands' location on the Atlantic Flyway, the district gets a steady stream of birds most times of year. Even in winter, typically a slow time for watching birds in the Northeast, the landfills and marshes still get amazing numbers of birds from the boreal forest of Canada and other points north – including Rough-legged Hawks and all manner of ducks.

Significantly, a region that was once on the endangered list and pretty much given up for dead now is home to plenty of birds considered threatened, endangered or of special concern by the state of New Jersey. In some instances, birds that were rarely seen are now frequently fliers – notably such raptors as the Bald Eagle, the Peregrine Falcon, the Northern Harrier, and (from mid-March to early November) the Osprey.

Endangered Peregrine Falcons now nest on several man-made structures in the Meadowlands.

Yellowlegs take wing in the Saw Mill Creek Wildlife Management Area as night falls.

In all, at least 16 birds on the state's watch lists nest in the Meadowlands, and the Bald Eagle has successfully nested just north of the district.

- **Four endangered species:** Pied-billed Grebe, Peregrine Falcon, Least Tern and Northern Harrier.
- **Five threatened species:** American Kestrel, Osprey, Savannah Sparrow, Yellow-crowned Night Heron, and Black-crowned Night Heron.
- **Seven species of special concern:** Barn Owl, Brown Thrasher, Common Nighthawk, Least Bittern, Saltmarsh Sparrow, Snowy Egret, and Spotted Sandpiper.

Of the lot, the two most elusive and mysterious are the two species that hunt at night – the Barn Owl and the Common Nighthawk.

Although many bird-watchers long suspected that the nighthawks nested on the closed landfills, confirmation arrived only recently when a construction crew that was remediating the old Kingsland Landfill in Lyndhurst notified the Meadowlands Commission that they had found an injured Osprey in an area they were bulldozing.

Pied-billed Grebe, an endangered species in New Jersey, nests in the Kearny Marsh.
Photo by Ron Shields.

Least Bitterns, a secretive bird of special concern in New Jersey, breed in the Kearny Marsh.
Photo by Ron Shields.

NJMC Naturalist Mike Newhouse investigated, expecting to transport the injured bird to a raptor rehabilitation center. But instead of finding an Osprey, he saw a Common Nighthawk on the ground. Nearby were two baby nighthawks, covered with feathers but too young to fly.

A construction worker called out that another bird was sitting on the ground about 20 yards away: another nighthawk, sitting on eggs. Calling the nighthawk's incubating place a "nest" would be an overstatement. When the bird flew briefly, she revealed two small speckled eggs in a tiny depression on the dirt-and-gravel surface.

When Marcia Karrow, Executive Director of the Meadowlands Commission, heard the news, she asked Newhouse to cordon off the nesting site, and told the contractors to stop work near the nesting site until all the young had fledged. About a week later, the two babies flew for the first time. Sadly, a landfill predator (likely a coyote or fox) took the two eggs a few days later, but the point was clear: Protected species were not to be taken lightly.

Common Nighthawks, another bird of special concern in New Jersey, nest on closed landfills in the district.

In the early summer of 2011, workers stumbled upon a female Common Nighthawk sitting on two eggs. The male was tending to two young Common Nighthawks 20 yards away.

The other nocturnal birds threatened by severe habitat loss are Barn Owls, which have nested in the Meadowlands in several vacant buildings and a trash-baling facility for years. The closed landfills offer perfect undisturbed hunting grounds. The charismatic raptors do the best to keep their nests a secret, and the Meadowlands Commission as a matter of policy does not disclose the nest locations. (The Barn Owl photo was taken during a recent Christmas Bird Count, after the owl flew out of an empty building.)

In recent years, the Meadowlands has played a host to many other birds seen infrequently in the region, including a Northern Wheatear, an American Avocet, a Northern Shrike, a Black-necked Stilt, and a pair of Snowy Owls. The wheatear and the stilt stayed less than a week, but both the shrike and Snowy Owls hunted the old landfills all winter long, and the avocet hung around Harrier Meadow for a month.

Elusive Barn Owls nest in several empty buildings in the district.

Two Snowy Owls wintered in Lyndhurst during the winter of 2008-09. This Snowy Owl liked to perch atop a muskrat hut.
Photo by Herb Houghton.

A Northern Shrike arrived along Disposal Road in mid-December 2009, and stayed through St. Patrick's Day, 2010.
Photo by Ron Shields.

This rare (for the region) Northern Wheatear spent several days along the Transco Trail in DeKorte Park in September – and attracted more than 100 birders – before continuing its migration to Africa.

American Avocets occasionally hang out in the region. The one pictured above stayed in DeKorte Park for a few days in 2010. Another stayed in Harrier Meadow for several weeks in 2011. *Photo by Jeff Nicol.*

Black-necked Stilts, infrequent visitors to the region, sometimes visit Harrier Meadow in North Arlington.

The two-lane road was built for one purpose: to give trash-haulers access to the Erie and Kingsland Landfills. The landfills have long been closed, and the roadway is pocked with potholes and plenty of unmarked speed bumps, but birders typically find it well worth the bone-jarring trip.

Consider: on one October Saturday a year or two ago, photographer Ron Shields reported that he had just seen and taken pictures of "distant Adult Bald Eagle, multiple flyover Ospreys with catch, numerous Northern Harriers (three on one occasion), Merlin with catch, Am. Kestrels hunting in teams of two, Red-tailed Hawks, and a Peregrine Falcon."

What makes Disposal Road such a draw is that birders and photographers are able to get great looks at the raptors and rarities for minutes at a time as they hunt the landfill or perch there.

Other raptors seen from Disposal Road include Bald Eagle, Turkey Vulture, Black Vulture, Great Horned Owl, Red-shouldered Hawk, Osprey, Sharp-shinned Hawk, and Cooper's Hawk.

Come late December, the landfill often attracts rare Rough-legged Hawks, which breed on the Arctic tundra and head south when they can't find enough prey. The birds, cousins of the Red-tailed and Red-shouldered Hawks, come in two varieties, dark morph and light morph.

Perhaps the biggest attraction of all was the Northern Shrike, which lived on the Kingsland Landfill from December 2009 to March 2010, impaling its prey on a nearby barbed-wire fence or on trees with thorns.

If you want to see raptors and rarities up-close in the fall and winter, few places in the entire Northeast have the lure of a stretch of macadam with the unlikely name of Disposal Road.

Once used by garbage trucks en route to adjacent landfills, Disposal Road now attracts raptors, birders, and photographers each winter.

Gray Ghosts – male Northern Harriers – hunt above Disposal Road's old landfills in winter. *Photo by Ron Shields.*

A female Northern Harrier looks for prey on the old Kingsland Landfill. *Photo by Ron Shields.*

Bald Eagles have made an incredible comeback and are seen often throughout the region. *Photo by Ron Shields.*

Rough-legged Hawks arrive along Disposal Road in winter to hunt the old landfills. *Photo by Ron Shields.*

What Disposal Road is for raptors in winter, DeKorte Park is for ducks – to the point where the Meadowlands Commission has printed a locator map that lists 14 species that birders might see wintering in the tidal impoundments. The ducks migrate from as far away as the western plains, Canada, and even Alaska to winter in the Meadowlands.

In the spring, summer, and early fall, DeKorte and several sites along the Hackensack River are also great places to view migratory shorebirds, including sandpipers by the thousand.

The Goldeneye makes an occasional appearance in DeKorte Park's tidal impoundments in winter. *Photo by Ron Shields.*

Buffleheads winter by the dozen in DeKorte Park and elsewhere in the region. *Photo by Ron Shields.*

Birders looking for hard-to-see waterbirds and an assortment of ducks and raptors gravitate toward the fresh-water Kearny Marsh, where unusual birds abound – including American Coots, uncommon Common Gallinules, Black Skimmers, Least Bitterns, American Bitterns, Glossy Ibises, and Merlins. Many of these birds are best seen from a gently paddled kayak.

American Coots – which sport red eyes, silver bills, and lobed feet – congregate in Kearny Marsh in winter. *Photo by Ron Shields.*

Common Gallinules nest in the Kearny Marsh and DeKorte Park. *Photo by Ron Shields.*

A Black-crowned Night Heron chows down on a Blue-clawed Crab in the Kearny Marsh. *Photo by Ron Shields.*

Merlins – medium-size falcons – are being seen more often these days in the Kearny Marsh and elsewhere in the district. *Photo by Ron Shields.*

Glossy Ibises occasionally stop in the Meadowlands for a day or two in the spring or summer. *Photo by Ron Shields.*

Seldom-seen Blue-winged Teal take flight in the Kearny Marsh. *Photo by Ron Shields.*

Black Skimmers fly in formation in the Kearny Marsh. *Photo by Ron Shields.*

Chapter Six
Of Muskrats and Other Mammals

I have always found small mammals enough like ourselves to feel that I could understand what their lives would be like, and yet different enough to make it a sort of adventure and exploration to see what they were doing.
– Donald R. Griffin, *Echoes of Bats and Men*

Wildlife surveys of the Meadowlands have recorded more than 25 species of mammals. Aside from humans and their pets, feral cats, Eastern Cottontail Rabbits, and Eastern Gray Squirrels, the vast majority are seldom seen.

While you might spot a Common Muskrat paddling along in a marsh, or a woodchuck popping its head out of its burrow, the landscape is a far cry from 60 years ago, when pigs outnumbered people in Secaucus and the muskrat population once outnumbered the human population in parts of the Meadowlands.

In the 1950s and 1960s, local hardware stores set up sales displays each winter for muskrat traps, and kids earned money trapping the animals instead of delivering the local daily newspaper.

As recently as the mid-1980s, muskrats abounded in the local marshes. Some two dozen trappers supplemented their incomes by catching between 10,000 and 15,000 of the foot-long critters during the winter months. Pelts, bound for overseas markets, went from $6 to $10 each.

Since then, some of the muskrats' prime habitats, dense Phragmites marshes, have been replaced with native wetlands plants and tidal flats. That has benefited migratory shorebirds, wading birds, and waterfowl at the expense of the muskrat population, but you can still see their lodges amid the Phragmites in the Saw Mill Creek Wildlife Management Area and other marshes throughout the Meadowlands.

These days, most of the district's wild mammals are small enough to go undetected (Masked Shrew, Eastern Mole, Meadow Vole, and three species of mice), or they are shrewd enough to avoid human contact. Many of the larger wild mammals live on closed landfills or in remote wetlands, and many hunt by night. Norway Rats likely hang out around a garbage-baling facility in North Arlington, where they are part of the food supply for Barn Owls and other raptors.

For better or worse, Virginia Opossums, Striped Skunks, and Northern Raccoons are visible mostly when they become road kill, and the most conclusive proof that River Otters live in the region was the discovery of a dead one sprawled on the shoulder of the Belleville Turnpike in Kearny. Until then, signs of otter activity were limited to occasional footprints.

A Groundhog heads for the marsh just before the start of a guided walk at DeKorte park.

Muskrats can still be seen regularly in the region, though their numbers appear to be in decline.

Jack Churuti of Jersey City, trapper in the meadows for some 50 years, chops through six inches of ice to pull up muskrat. Pelts bring 75 cents apiece. Churuti traps 750-800 a season.

This photo of a long-time muskrat trapper accompanied a 1959 *Newark Evening News* series on the Meadowlands. He trapped 800 to 900 Muskrats each winter – getting 75 cents for each pelt. *N.J. Meadowlands Commission archives.*

The Meadowlands Commission also gets reports of Red Foxes from time to time – often on a closed landfill. A family of foxes had a den just outside DeKorte Park in the early 2000s, a Red Fox was photographed at Mill Creek Marsh in 2008, a pair of bedraggled foxes lived near a radio transmitter in Carlstadt in 2010, and another pair was seen in Harrier Meadow in the fall of 2011.

Every so often, the commission receives reports of seals sunning themselves on docks along the Hackensack, sometimes accompanied by grainy cell phone photos. The seals are mostly Harbor Seals, with one exception. In early 2011 Stephen McNamara, operations manager with the Dawson Corporation, was inspecting a wetlands-remediation project at the Richard P. Kane Natural Area in Carlstadt, when he saw a five-foot-long Harp Seal sunning itself. The seal is typically seen in the Arctic Ocean and Northern Atlantic – without the cell phone photo, the sighting would have gone undocumented.

Red Foxes live near marshes, in natural areas, and on old landfills throughout the region. This one was photographed at Mill Creek Marsh. *Photo by Sanford M. Sorkin.*

A rare Harp Seal surfaced in the Richard P. Kane Natural Area one winter a few years back. A cell phone photo documented the appearance. Harbor seals are seen more frequently but are still uncommon. *Photo by Stephen D. McNamara.*

Eastern Coyotes also do their best to avoid humans, though motorists' headlights sometimes cast them in a harsh light as they cross the road near landfills near dawn or dusk. By the time you can train a camera on them, they are long gone. But just as surveillance cameras reveal all sorts of human activities, a digital camera with a motion-sensor shutter release can provide a window on an otherwise unseen landscape.

Kenneth DeMatteo of Central Jersey Pheasants Forever placed one of these "trap" cameras on the closed 1-E Landfill in North Arlington for several months and caught several coyotes unawares, as well as Eastern Cottontails, skunks, White-tailed Deer, Red Fox, raccoons, muskrats, and Snapping Turtles). DeMatteo estimates that more than a half-dozen coyotes live on the four-hundred-acre landfill.

A motion-sensitive camera on the old 1-E Landfill captured this image of an Eastern Coyote. *Courtesy of Kenneth A. DeMatteo of Central Jersey Pheasants Forever.*

White-tailed Deer are slowly making inroads into the region.

The most elusive mammal has been the American Mink. Dr. Ross Feltes, Supervisor of Natural Resources Management for the Meadowlands Commission, says that a few sightings have been reported but they were never substantiated.

Similarly, proof of bat activity in the Meadowlands was mostly anecdotal until the summer of 2011. People reported seeing bats just after dusk but didn't – or couldn't – photograph them, and no roosts have been reported in the district.

With the help of the Conserve Wildlife Foundation (CWF) of New Jersey and the state Division of Fish and Wildlife, the Meadowlands Commission conducted an acoustic bat survey covering parts of Kearny, North Arlington, Lyndhurst, and Rutherford. The tool of choice: an AnaBat SD2 acoustic detector.

The device is essentially a large, super-sensitive microphone that can record the ultrasonic call patterns of bats. Bats navigate the skies through echolocation – constantly emitting high-frequency pulses of sound that bounce off nearby objects. The bats' ears hear these returning pulses as echoes, providing information about the size, shape, and whereabouts of the objects.

According to MacKenzie Hall of the CWF, "For the most part, bat calls are ultrasonic – meaning they're above the 20 kHz upper range of human hearing. So we can't hear them…but an acoustic detector can."

CWF's AnaBat acoustic detector microphone can detect bats echolocating from up to 300 feet away, and a memory chip stores each bat call for later analysis.

"Since different bat species have unique call patterns that can be used to tell them apart, acoustic surveys give us an idea of the abundance and diversity of bats in an area of interest," she says.

One June evening in 2011, just past sunset, a CWF staffer strapped an AnaBat SD2 to the roof of an SUV and drove through four district towns at fifteen miles an hour. The device recorded a total of 24 bat calls in roughly 45 minutes – 22 Big Brown Bats and 2 Red Bats. A third call pattern – which appears to have been made by a Hoary Bat – was detected as well, but it contained just a single pulse of sound and could not be confirmed.

The test drive proved that the Meadowlands has a population of bats, and that's good for a lot of reasons.

Says CWF's MacKenzie Hall, "The Meadowlands are loaded with marshes, rivers, and tidal wetlands. Bats should have plenty to eat and drink here, so it's great to learn they're taking advantage. In return, they munch down the mosquitoes, Tent Caterpillar moths, Potato Beetles, flies, and other pests for us. Without bats, we'd either have skies full of bugs or an environment full of pesticides."

This high-tech, super-sensitive microphone records the ultrasonic call patterns of bats. Researchers conduct bat censuses by strapping the device to the roof of a vehicle and driving slowly through an area just after dusk.

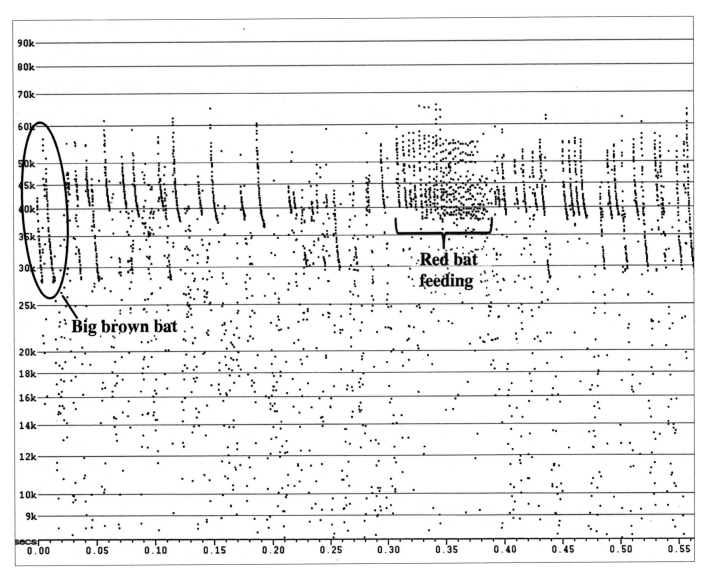

The AnaBat acoustic detector microphone can detect bats echolocating from up to 300 feet away. This sonogram shows two species echolating – a Red Bat feeding and a Big Brown Bat as it flew. *Courtesy of MacKenzie Hall.*

Chapter Seven
New Uses for Old Landfills

There is nothing in which the birds differ more from man than the way in which they can build and yet leave a landscape as it was before.
— Robert Lynd, *The Blue Lion and Other Essays*

Who would have guessed it? Landfills, once enormous open sores on the Meadowlands' landscape, now represent prized open space. After hundreds of millions of dollars in clean-up efforts, the landfills no longer pose the dangers of yore, and they constitute some of the last large parcels of vacant land in the metropolitan region.

While some of the low-elevation former dumps are slated for redevelopment, the taller landfills – the ones that now resemble huge hills – will become solar farms, parks, and habitat for migratory birds. The old landfills are already used to generate electricity. Since 1989, roughly a billion cubic feet of methane a year have been extracted from the landfills and converted to enough electricity to supply some 10,000 to 11,000 homes annually.

The 1-A project uses 12,500 panels to convert energy from the sun into electricity.

Sunflowers on the sides of old landfills provide a food source for migratory birds.

Adam Levy, the NJMC attorney who has been leading the agency's solar initiative, says that such factors as environmental concerns and still-decomposing contents in these larger landfills create challenges for commercial, industrial, or residential development.

"The perfect use for these landfills is something that's static and that can take advantage of available open space," says Levy. "In a densely developed place like this, where open space is so valuable, you're never going to find another spot where you can install solar panels without someone finding a more beneficial use for the land."

In a partnership with SunDurance Energy and PSE&G, the commission has created a 3-megawatt, grid-connected solar farm atop the 1-A Landfill in Kearny. The once-orphaned 35-acre landfill is now home to an $18 million project that features 12,500 solar panels. Long-term, the commission hopes to build similar arrays on other large former dumps.

"Wind turbines and other complex renewable-energy systems are very effective in certain environments, but they are huge mechanical structures that can fail or topple over, and they have environmental issues like their impacts on migratory birds," says Levy. "Solar is a super simple technology that requires relatively little maintenance or monitoring. It is essentially a solid surface that generates electricity when exposed to the sun. How can you get simpler than that? The efficiency is only going to improve, and as long as you have the space for it, it's a no-brainer."

As of late 2011, New Jersey accounted for 24 percent of the nation's solar arrays. The long-term goals of the solar initiative are simple as well – the generation of clean energy, reduced dependence on fossil fuels, reduced carbon footprint, and reduced energy costs.

The old 1-A Landfill in Kearny can generate 3 megawatts of electricity at any given moment – and 3.6 million kilowatt-hours annually.
Courtesy of the Meadowlands Environmental Research Institute.

But one old dump is being put to an even more basic use. The 42-acre former Erie Landfill in North Arlington has become an open-air laboratory for bird research. Each spring and fall since 2008, NJMC naturalist Mike Newhouse and his band of Ramapo College students and volunteers have placed mist nets in the various habitats growing on and near the landfill – which range from stands of Black Locust trees to an invasive species called Mugwort – to determine how songbirds are using these places.

The long-term goal is to take advantage of some of the open space on closed landfills. One opportunity for such open space would be to plant seeds of grasses that attract migratory songbirds as well as such threatened and endangered species as Grasshopper Sparrow and Bobolink.

The researchers open the nets each weekday just after dawn and wait for birds to fly in. Once the birds are caught, the volunteers place them in soft cotton bags and take them to a makeshift bird-banding station where the birds are weighed, measured, and banded with a small aluminum tag that includes a unique serial number for future reference.

Newhouse and his crew have banded and released more than 17,000 warblers, sparrows, and other small birds thus far. The number of species has totaled more than 100. One of the nice surprises: The Savannah Sparrow, a threatened species in New Jersey, has been one of the most common species captured – more than 4,000 in three years.

"We want to see how migratory birds are using the habitats on a closed landfill, and how we can manage those habitats to help migratory birds," says Newhouse. "We keep track of which habitat we catch each bird in, and that way we can determine which habitats are the most productive for species diversity and abundance."

Among the findings: The birds do not necessarily like stands of Black Locust trees – there's no real food in there for them. Some small stands of Cottonwood trees do bring in diversity, and a lot of the woodland species that are migrating will stop in these patches for a little bit. Most of all, birds like a variety of plants that produce different seeds – like Foxtail Grass, Mugwort, and Sunflower seeds – and not a monoculture of Mugwort or Phragmites.

The team bands the birds for two main reasons: The U.S. Geological Study uses the data for its migration and longevity research, and the Meadowlands Commission uses the data for its study of stop-over ecology. When birds migrate long distances, they stop for a few days at certain areas to regain their fat reserves while they wait for the next cold front or warm front to push them toward their breeding grounds or wintering grounds.

"We're trying to figure out how the Meadowlands helps birds as they migrate," says Newhouse. "We'll catch a bird for the first time and band it with that unique number. If we catch the same bird at a later date and take the same measurements, we can see if the bird is gaining fat reserves to continue their migration. We have seen this with our recaptures. A lot of them are gaining weight."

To help fund the research, Newhouse has organized a team to compete in New Jersey Audubon's World Series of Birding, which placed second in its division in 2011, spotting 127 species in less than 24 hours.

Researchers use mist nets on the sides of the old Erie Landfill to briefly capture migratory birds in the spring and fall.

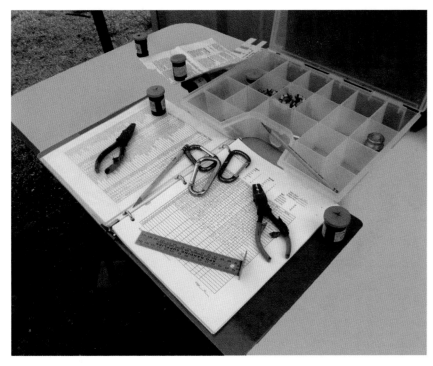

Each bird is weighed, measured, aged, and banded with an individually numbered aluminum band. That information is recorded for inclusion in a national database.

This Palm Warbler is among the more than 17,000 small birds banded by the NJMC in recent years.

A banded American Goldfinch picks at some thistle along
Disposal Road. The species is the state bird of New Jersey.
Photo by Ron Shields.

The Erie Landfill has also been the site of a high-tech bird study, done in conjunction with the vaunted Cornell Lab of Ornithology.

"We have a microphone on the top of the landfill, which is recording the sounds made by birds migrating at night," says Newhouse. "The microphone is owned by the Cornell Lab of Ornithology, and they plug that recording data into a software program that can identify the species and count how many of each species."

Newhouse is taking that data and comparing it to the data collected at the banding station to see what species are flying over and what species are staying over in that part of the Meadowlands – and what habitat they are using.

Although the data was still being analyzed at the time this book went to press, Newhouse has already learned quite a bit from the microphone's recordings: "We do know that there are tens of thousands of birds that fly over at night – birds that we would not know about without the microphone."

Two American Kestrels perch atop the Cornell Lab of Ornithology's microphone on the old Erie Landfill.

Chapter Eight
Butterflies and Bugs

Two-legged creatures we are supposed to love as we love ourselves. The four-legged, also, can come to seem pretty important. But six legs are too many from the human standpoint.

– Joseph W. Krutch

"Delicate" is a word that typically does not come to mind when one thinks of the Meadowlands, a world of marshes, mudflats, highways, railroads, bridges, and buildings of all sorts.

But if you work your way down the food chain to insects, you will see some of Mother Nature's most delicate creations – butterflies and other fascinating insects – surviving and even thriving in this urban environment.

A Monarch perches on a Pokeberry. *Photo by Marco Van Brabant.*

A Painted Lady nectars on a Purple Coneflower.

Aside from their beauty and their role as pollinators, butterflies are important because they are a wonderful way to introduce children and adults to nature. They don't sting, bite, buzz, eat wool clothing, or infest your pantry. And since butterflies are most often seen near flowers in warm, sunny weather, they are associated with happy times, and happy memories.

"Butterflies are ephemeral, graceful creatures that people connect with in a magical way," says Jeffrey Glassberg, author of *Butterflies through Binoculars*.

Although some naturalists like to point out that many butterflies are anything but dainty – Monarchs, after all, can migrate more than 1,000 miles – everything is relative.

An Eastern Tiger Swallowtail nectars on a Butterfly Bush.

"Butterflies as a group are an important indicator species because they are exquisitely sensitive small creatures in the environment," Glassberg says. "They are balanced on a knife's edge. Unlike birds or mammals, which tend to have slow declines, butterflies' populations can fall off the cliff or shoot up to the sky based on small changes in the environment."

The Meadowlands is home to some two dozen species of butterflies, and the butterfly list at DeKorte Park keeps expanding as more and more butterfly-friendly plants are added to the landscape. What's more, as increasing numbers of butterfly watchers visit the park's gardens and Butterfly Bushes, they find species that had previously flown under the radar, so to speak.

A butterfly egg is a tiny dot on a Common Milkweed leaf.

This caterpillar will eventually become a Monarch.

Black Swallowtail.

In the past few years alone, such show-stopping butterflies as the Variegated Fritillary and the Pipevine Swallowtail have been added to a list that already included such mainstays as Monarch, Viceroy, Common Buckeye, Eastern and Black Swallowtail, Pearl Crescent, Painted Lady, and a raft of skippers.

"Over time, the more diverse the habitat in the Meadowlands becomes, the even better it will be for butterflies," says Glassberg.

A great place to see a wide variety of species is the Jill Ann Ziemkiewicz Butterfly Garden, located just outside the Environment Center. The 50-foot by 80-foot garden was built to honor the memory of a 23-year-old Rutherford resident, the youngest member of the flight crew on TWA Flight 800, which crashed into the ocean off Long Island in 1996.

The fountain in the garden's center is in the shape of a sunflower, Jill's favorite. Every year around the anniversary of the crash, her loved ones leave sunflowers at the garden in her memory.

One of the challenges – and great rewards – of watching butterflies is finding butterflies. After all, some of the tinier species, such as Eastern Tailed-Blues, are about the size of a dime. To find these tiny marvels, you really have to pay attention to the world around you – which in turn enables you to notice other amazing insects.

The Variegated Fritillary was added to the DeKorte Park butterfly list in recent years, along with …

...the Pipevine Swallowtail. Increased plant diversity helps attract more species.

Take the Clear-winged Moth, an incredibly nifty bug that was seldom noticed and mostly unheralded in the Meadowlands until more and more nature lovers and photographers started flocking to DeKorte Park and witnessing this bug for themselves.

These moths, often mistaken for bees or tiny hummingbirds, come in two varieties at DeKorte – the Hummingbird Clear-winged Moth and the Snowberry Clear-winged Moth (easiest way to tell them apart is that the former's thorax is more reddish and the latter's is blackish).

The moths zip back and forth between Butterfly Bushes and other flowering plants and sip nectar through their proboscis, which amounts to a straw. Their "clear" (scaleless) wings enable you to still see the flower as they feed. Unlike many moths, these little dynamos feed and fly during the daylight hours.

Clear-winged moths are attracted to the Butterfly Bushes in the summer months.

Two insects that typically fly undetected in the Meadowlands are dragonflies and damselflies – seen mostly along the edge of marshes in the summer months. On hot sunny days, DeKorte Park's Marsh Discovery Trail and the Kearny Marsh are good spots to look for both species.

Erik Kiviat's 2007 study, "Monitoring Biological Diversity in the Hackensack Meadowlands," found 14 species of dragonflies and 10 species of damselflies – not including the common Green Darner dragonfly.

"Much of the Meadowlands is brackish and favors a small suite of species already present – such as Needham's Skimmer, Seaside Dragonlet, and Big Bluet – while naturally excluding many others," says Allen Barlow, author of *Field Guide to Dragonflies and Damselflies of New Jersey*. "I would expect a slight increase in diversity as the habitat improves."

Three common dragonfly species are the Green Darner (above), the Seaside Dragonlet (right) and the Black Saddlebags (page 92 top).

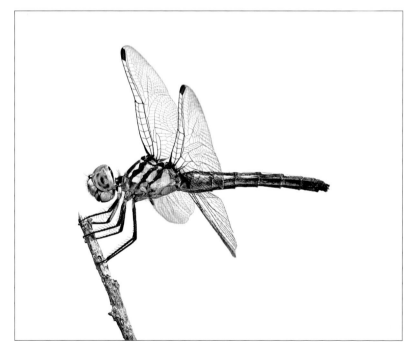

The Seaside Dragonlet. *Photo by Marco Van Brabant.*

The Black Saddlebags.

This bluet is one of the common damselflies in the region.

Some bugs are anything but dainty. Take cicadas, the thick, green bugs that arrive each May and create an annoying chorus of buzzes. Fortunately, DeKorte Park also attracts the aptly named Cicada Killer, which arrives a few weeks later.

This large wasp has an ingenious reproductive strategy. The female attacks and paralyzes a cicada, drags it to her burrow, lays an egg on the prey, and then seals off the burrow. A grub hatches in a few days, eats the cicada, weaves a cocoon and overwinters underground. The insect pupates in the spring, digs its way to the surface, and begins the cycle all over again.

Praying Mantises are typically seen in late summer in DeKorte Park and Harrier Meadow. Although these striking bugs tend to blend in well with their surroundings, they have proven to be a nice snack for American Kestrels, which eat them on the fly.

For insect diversity in the Meadowlands, it's tough to top bees. A study by New Jersey Institute of Technology researchers a few years back documented 78 species of bees in the district, including two that had never been seen before in North America – a European Thistle Bee and a North African bee with a red tail and yellow face. The two species were likely stowaways in shipping pallets or crates that arrived in the region as a result of global trade.

The Meadowlands Environmental Research Institute commissioned the study to see if the district was providing enough habitat for native pollinators, so they could help compensate for the recent decline in honeybees as a result of colony collapse.

The Cicada Killer Wasp has an unusual, if gruesome, reproduction strategy.

A Syrphid Fly – often mistaken for a type of bee – works a Threadleaf Coreopsis bloom in Jill's Butterfly Garden.

A Praying Mantis provides an in-flight meal for an American Kestrel.

marsh mosquitoes used to fly up Broad Street from the marsh right into downtown Newark. But those days are past."

The reason, according to Kent, is that breeding salt marsh mosquitoes prefer stagnant, shallow, and often polluted water. As a result of all the efforts in the Meadowlands to open marshes to tidal flow and to bring wildlife back – including all those bug-zapping Tree Swallows – there's not as much stagnant water around.

"The mosquito problem is still bad in some parts of the state," says Kent, "but in the Meadowlands it's not nearly as bad as it was."

Midges, ubiquitous in DeKorte Park in late spring, have feathery antennae that can bug people.

Some Meadowlands insects, of course, you can often encounter without going looking for them: midges and mosquitoes.

Midges abound in early May. They don't bite, but can be extremely annoying if you happen to walk too close to a bush where they are swarming. One of the reasons they are literally irritating – their heads sport long feathery antennae.

No chapter on insects would be complete without a mention of the mosquito, which was once famously described by Andy Warhol as "the state bird of New Jersey."

In the late 1800s and early 1900s, the Meadowlands was considered the mosquito capital of the world.

"Historically, mosquito control as a science and as a practice got its start in that neck of the woods due to the nature of the unmanaged meadows," says Robert Kent, the administrator of New Jersey's Office of Mosquito Control Coordination. "The salt marsh mosquito was very prolific in the marshes around Essex County, Union County, Bergen County and that neck of the woods. My grandmother, who lived in downtown Newark, recalled rubbing citronella oil all over herself because the salt

Chapter Nine
A Meadowlands Year

To be interested in the changing seasons is a happier state of mind than to be hopelessly in love with spring.
– George Santayana

The Meadowlands is a world of extremes, a world where the summers seem hotter than the surrounding areas, the winters seem colder, and the winds stronger.

Aside from Laurel Hill and the old landfills, the region is as flat as a North Arlington pizza. In the summer, trees are too few and far between to provide much shade from an often relentless sun. In the winter, the winds come barreling across the open water, marshes and mudflats.

The Meadowlands is also ruled by the Hackensack River's tides, a twice-daily ebb and flow of as much as seven feet of water that can turn vast expanses into mudflats and reduce tributaries to bare creek beds.

To understand the Meadowlands, one has to grasp this dynamic place in all seasons.

Tides cover the region's mudflats with several feet of water twice a day, drastically changing the landscape. *Photo by Marco Van Brabant.*

January

The Meadowlands can be a frigid place this time of year, with wind and cold providing a double-whammy for any human who braves the outdoors for extended periods. The thousands of ducks that over-winter in the region's tidal impoundments and other open water seem to have the right approach to the weather. The gracefully sculpted Canvasbacks and smaller Green-winged Teal and Ruddies congregate in rafts that number well into the hundreds, tucking their heads on their chests, and chilling. Other species tend to hang out in lower numbers, ranging from dozens of Northern Shovelers to a few Buffleheads and an occasional Goldeneye. The ducks mostly stay in the channels at low tide, aside from the Teal, which like to snarf up anything edible as they waddle through the nearby exposed mud.

After a snowfall, the open spaces of the district are especially enticing because long stretches of trails remain untrammeled by human footprints, even though they are in the midst of a metropolitan area of more than 12 million people.

Out-of-the-way places like Skeetkill Creek Park in Ridgefield and Losen Slote Creek Park in Little Ferry seem altogether different after a snowstorm. Both locales, located on the doorstep of residential or commercial development, often seem devoid of bird life. With a cover of snow and with trees barren of their leaves, civilization seems much closer.

(Right and following page) After a snowstorm, high winds and sub-freezing temperatures alter the landscape. *Photos by Marco Van Brabant.*

A Northern Shoveler paddles about in Teal Pool, one of DeKorte Park's tidal impoundments. *Photo by Ron Shields.*

February

Once you get away from the developed areas, the Meadowlands is still a barren place, save for the fact that the indelible mark of humankind looms wherever you turn – the old landfills filling the horizon, the endless ropes of telephone wires and power-transmission lines, and the gigantic AM-radio broadcast towers jutting 250 feet and more into the sky.

The top of a former landfill in North Arlington is wind-swept tundra, where Snow Buntings and Horned Larks alternately flatten themselves against a patch of bare ground or scour their surroundings in search of their next meal.

A visit to a frozen Mill Creek Marsh is a reminder of how big the Meadowlands' sky can be. Against a seemingly endless bright blue sky, an army of cumulus clouds plods eastward. A lone Northern Harrier crisscrosses its way across the marshes and patches of open water, ready to strafe any unsuspecting ducks.

Late in the month, an overnight snowfall turns into a cold rain, and the typical Meadowlands vista – marsh, ice, and leaden sky – has the texture and hues of an Andrew Wyeth watercolor.

The Rough-legged Hawks have high-tailed it northward, and the rafts of Canvasbacks have begun to leave for their northern nesting grounds. A Long-eared Owl that roosted clandestinely next to DeKorte Park's main parking lot for much of the winter has disappeared as mysteriously as it arrived.

Much of the region becomes a no man's land in winter, crisscrossed by utility lines, railroad tracks, and highways. *Photo by Marco Van Brabant.*

Horned Larks congregate on the tundra-like surfaces of the old 1-E Landfill.

In many places, the first robin is supposed to be the harbinger of the season, but in the Meadows, it's the quirky Killdeer, who acts as though it owns the entire region, zipping around making its annoying dentist's-drill call and laying its eggs on any flat surface it finds, from gravel rooftops to ball fields to parking lots.

A Killdeer, one of the harbingers of spring, returns to Disposal Road in late winter. *Photo by Ron Shields.*

March

Except for a few lingering slate-gray icebergs created by snowplows, winter is in retreat. After an overnight rain, earthworms appear from nowhere and venture into the puddles.

Across the Meadowlands, male Red-winged Blackbirds begin to arrive in the marshes and promptly stake out their turf with a cacophony of calls. At Laurel Hill and just outside DeKorte Park just after sunset, Woodcock are performing their courtship displays. If you stand still and listen, you are instantly reminded how urban the Meadowlands' open spaces are, and how adept the human ear is to filtering out the endless urban background din.

The immediate noise is the rumble of traffic from the eastern spur of the New Jersey Turnpike a few hundred yards away. Then it is the roar of a jet plane headed for Newark. And then it is the rumble of a commuter train pulling into the Secaucus Junction Rail Station. But if you listen carefully, you can hear the March wind riffling through the dry Phragmites, and the distinctive buzzes of male Woodcocks during their courtship displays.

Tree Swallows start staking their claim to the hundreds of volunteer-built nesting boxes in late March.

By mid-month, a change of seasons is in the air – not just the warming breeze but the first of the Tree Swallows as well, as sure a sign of spring as setting the clocks an hour ahead.

Days later, the first of the Ospreys moves through. The Common Ravens are bringing nesting material to their home in the cliff at Laurel Hill again, and a smattering of Great Egrets is arriving. A few Great Cormorants still loiter on the swing bridge while the Double-crested Cormorants return.

At month's end, Wilson's Snipes, with their long sipping-straw bills and camouflage plumage, are feeding in Harrier Meadow, and the sun's warmth lingers into late afternoon.

A Wilson's Snipe blends in with shoreline in Harrier Meadow as it makes a pit stop in its northern migration.

April

The Meadowlands is becoming a huge open-air maternity ward. An NJMC staffer reports the first Killdeer nest sighting of the season in a nearby town, with four neatly camouflaged eggs sitting in a depression in a gravel parking lot. When humans approach, the female limps away as if on a broken wing.

Nearly every Tree Swallow box – and the district is home to at least 400 of them – gets a tenant as soon as the birds arrive from points south. In instances where the new housing is installed by boat, swallows fly up to a nest box's tiny opening even before the box is installed along the water's edge.

Male Ruddy Ducks have Tiffany-blue bills in breeding season.

The Tree Swallow nesting boxes are so popular that some birds can't wait for them to be installed.

103

Ospreys have begun to set up house again in the usual places along the Hackensack River, and some have taken to catching lunch in DeKorte's Shorebird Pool. They hover for several moments, crash into the water and – if they're lucky – emerge with a nice fat carp or perch.

More Snowy and Great Egrets arrive daily, along with the yellowlegs. The wading birds work the shallows for food, occasionally side by side.

The Forsythia and Daffodils burst into bloom, a preview of glorious spring days to come. The male American Goldfinch, New Jersey's state bird, is now as yellow as the Daffodils nearby. The trees are budding. The Bleeding Hearts throb in bright pink, and the male Ruddy Ducks sport Tiffany blue bills

By month's end, DeKorte Park is filled with bird life – American Robins, Northern Catbirds, Blue Jays, Yellow-rumps, Eastern Towhees. A Peregrine Falcon returns to a familiar perch atop one of the many electric transmission towers nearby.

The Bleeding Hearts in DeKorte Park a sure sign of spring.

May

The main peloton of migratory warblers is arriving across the district. On an early walk at Losen Slote, we counted nine species as colorful and exotic as their names: Blue-winged, Blackburnian, Nashville, Common Yellowthroat, Northern Parula, Yellow-rumped Ovenbird, Black-and-white, Black-throated Green.

Baltimore Orioles, bedecked in orange and black, are taking up residence throughout the district. Their elaborate woven nests typically remain hidden from view but their distinctive clarion calls are audible at a distance.

The carp are spawning like crazed miniature sharks in DeKorte's Shorebird Pool, the Kearny Marsh, and other impoundments. Their dagger-like fins cut through the water's surface in the shallows.

Midges by the thousands swarm near the buildings and shrubs – providing much annoyance for humans and lots of protein for many of the birds.

Throughout the Meadowlands, the Princess Trees (a.k.a. Empress Trees or Pawlownia), are sprouting beautiful lavender blooms – an invasive tree originally from Asia that, at least for a week or so, goes from disliked or ignored to enchanted.

On a lunchtime walk, human beings are scarce on a rainy Marsh Discovery Trail, but it is far from deserted. A Spotted Sandpiper scurried down the boardwalk, pausing every so often to bob its tail feathers. A lone Black-crowned Night Heron perches on a railing just ahead, lost in whatever thoughts herons have. Above, the walkway is laced with swallows that zip back and forth.

Baltimore Orioles build their hanging nests near humans at DeKorte Park. This oriole was banded at the old Erie Landfill, just down Disposal Road.

Spawning Carp often resemble small sharks as they cruise the marshes and impoundments. *Photo by Ron Shields.*

June

The old stands of Phragmites are rapidly being overtaken by fresh green reeds, some already six feet high. Marsh Wrens abound throughout the region's wetlands. You can detect their presence mostly by ear – they have a haunting call that sounds like bamboo wind chimes in a breeze – but sometimes they perch high on the Phragmites to survey their surroundings.

Most of the shorebirds and warblers have continued on their journey north, but the butterflies are arriving as more and more flowers bloom. DeKorte Park looks like one giant bouquet of Oak Hydrangea, orange and yellow Daylilies, Cardinal Flowers, Coreopsis, and Butterfly Bushes.

Judging from their frequent calls, Orchard Orioles and Baltimore Orioles are nesting on DeKorte Park's Kingsland Overlook Trail and by the Carillon.

The days are growing so long that twilight still lingers at 9 p.m. Temperatures fluctuate between spring and summer, as if the seasons cannot make up their minds. A pheasant family – male, female, and nine little ones – makes an appearance along Disposal Road.

On the official first day of summer, Black Skimmers work the tidal impoundments and expanses of open water throughout the district, silently slicing the surface like winged bayonets.

At month's end, Tree Swallows are everywhere, including a perched youngster or two. At one point on the Marsh Discovery Trail, a male Red-winged Blackbird aggressively protects an unseen nest from unwary passersby, zipping down and snapping its wings at ear level – a jarring way to get a human's attention.

A Black Skimmer knifes through the surface of DeKorte's Shorebird Pool.

A banded Marsh Wren stakes out
its turf with a unique haunting call.
Photo by Ron Shields.

July

Come Independence Day, the Daylilies provide an endless fireworks display, just without the noise. Butterflies are starting to arrive in numbers, led by a few Monarchs. The Phragmites is going at full throttle and well over nine feet tall.

A few hundred yards beyond the district's northern boundary, in Ridgefield Park, a pair of Bald Eagles has successfully nested. Early one morning, two young eagles can be seen flying near the nest along Overpeck Creek. Only a few years ago, the eagles were occasional winter visitors. Nowadays they can be seen year-round, but they are still the biggest crowd-pleasers in the sky. A nesting site in the Meadowlands District is likely a matter of time.

Mid-month brings a heat wave, and the Meadowlands feels like an outdoor oven.

This is "Youngster Time" in the Meadowlands. At the Richard P. Kane Natural Area in Carlstadt, a distant blackish silhouette in the sky turns out to be a first-year Bald Eagle. In Harrier Meadow in North Arlington, a mother Gadwall and four trailing ducklings paddle together around one of the impoundments. At DeKorte Park, a baby Common Gallinule is found wandering near the Environment Center. The bird is so young that it still has spurs on the end of its wings to help it climb in and out of the nest.

All serve as a reminder about fragile life is, and how well it is faring in the marshes of the Meadowlands these days.

The Daylilies are running out of blooms, sporting bare brown stalks where blossoms once radiated. But as they fade, the Marsh Mallows, also known as Marsh Hibiscus, pop up in the marshes and along the river's edge. If the Meadowlands had an official flower, it would probably be the big, pink Marsh Mallow. This time of year, they seem to be everywhere.

In early summer, Daylilies help provide nature's fireworks. *Photo by Marco Van Brabant.*

Daylilies live up to their name – each flower blooms for only one day. *Photo by Marco Van Brabant.*

A Song Sparrow belts out its distinctive tune; Manhattan's Empire State Building is in the background. *Photo by Marco Van Brabant.*

A baby Common Gallinule looks for its mother along the marshy water's edge of the Shorebird Pool.

August

This month, the mudflats in the Meadowlands are king. The shorebirds start moving through in enormous numbers, looking for places to feed at low tide and places to roost when the tides are high.

One sandpiper in particular takes center stage. The bird is the tiny Semipalmated Sandpiper, weighing less than an ounce. Although New Jersey considers the bird of Special Concern, it abounds this time of year in the Meadowlands as it fattens up on its 4,000-mile migration from Canada to South America.

At low tide, these peeps swarm the region's mudflats, thousands of birds at a time, in search of the food they need to make the next leg of the trip south. At high tide at River Barge Park along the Hackensack River in Carlstadt, as many as 5,000 of the sandpipers cover the open dock space – unless an enterprising Peregrine Falcon dive-bombs them in search of a meal.

Meanwhile, many of the Meadowlands birds we have almost taken for granted earlier in the summer are harder to find. Many of the Orchard Orioles and Baltimore Orioles have either left or are packing their bags. The Red-winged Blackbirds are scarce, and the only Tree Swallows to be seen are migrants passing through from points north. Summer may still be going full blast, but some of the birds already sense it's time to go. The good news is that next February, when we are knee deep in winter, the swallows and Red-wings will be on their way north again. The seasons march on, often before we know it – but birds live in a parallel universe to ours, and their internal clocks know nothing of human time.

In DeKorte Park's Shorebird Pool, the egret population climbs, and the young Forster's Terns pound the water throughout the day. If you look at the Shorebird Pool's tide gates, you can see why so many birds are here: thousands of little Silversides, schooling by the inflow pipe.

In August and September, Great Egrets and Snowy Egrets by the dozen go fishing in DeKorte Park's Shorebird Pool when water levels are low.

Semipalmated Sandpipers migrate through the region by the thousands in mid- to late summer.

September

In 2011, a series of storms early in the month has pushed away summer and inundated the Meadowlands with rain and run-off. The cloudbursts have ruined several trails, dislodged a couple of marsh boardwalks, and generally rained down havoc on the region.

The storms serve as a reminder about how powerful Mother Nature is, and how vulnerable humans and their structures are. The wet weather has also brought bumper crops of mushrooms and mosquitoes. For all the jokes about the Meadowlands and mosquitoes, they aren't anything that some bug repellent won't solve.

The towering Phragmites sport purple tassels nowadays, and the Spartina has taken on a slight reddish glow. Sundown comes noticeably earlier and earlier. The vast majority of the peeps have vanished from the mudflats, their internal alarm clocks urging them southward.

A Northern Harrier patrols the old Kingsland Landfill near sunset. *Photo by Mike Girone.*

Nearing mid-month, the hawk migration picks up. When the winds are right – from the north or northwest – you can sometimes see raptors kettling above the North Arlington ridge, their wings spread and tail feathers fanned out as they gain altitude, then peel off and glide south. One warm sunny morning brings an avian parade – first several American Kestrels, then Sharp-shinned Hawks, and, every few minutes, a high-soaring Osprey.

Over the wetlands and landfills, you start to see Northern Harriers again, just one or two, as they zigzag across their hunting grounds.

Along the Marsh Discovery Trail, silence reigns in the late afternoon – no enchanting trill from the Marsh Wrens, no weird buzzing noises from the Forster's Terns.

A late migrating Common Buckeye stops by Jill's Butterfly Garden. *Photo by Regina T. Geoghan.*

October

Aside from an endless trickle of migrating Monarchs, some migrating Common Buckeyes, and a few Sulphurs and Cabbage Whites, butterflies are increasingly scarce across the Meadowlands. The Empress Trees have shed their leaves already, now lying dirt-brown and shriveled alongside the marshes. At Laurel Hill in Secaucus, the trees along the river flash their first hints of gold, copper, and silver. The season is changing, some days more perceptibly than others.

With the bright new hues of the foliage comes another changing of the guard. The warblers and shorebirds are on the wane, and the ducks are arriving in greater numbers. The ducks make for the better viewing. What they lack in plumage and song they make up for in sheer size, and they tend to swim in open water instead of perching in a thicket of leaves.

Other wintering marsh birds appear as well. American Coots and Pied-billed Grebes can be seen paddling about in the Kearny Marsh.

About mid-month, the Northern Pintails have begun to check in throughout the region. By December, they are so abundant in so many locations that, like the Semipalmated Sandpipers of summer, they are taken for granted. Appearances deceive. The pintail's numbers have plummeted by 77 percent over the past few decades. The tidal impoundments of DeKorte provide the perfect habitat for them to over-winter.

Spartina takes on a reddish patina in autumn. This scene is from Mill Creek Marsh. *Photo by Regina T. Geoghan.*

November

It's half-past autumn, and the days are bittersweet – Indian Summer days and cold nights under clear skies. The change of seasons is complete. Most of the trees are devoid of leaves, and the oriole nests and robin nests that had been hidden by foliage are now in plain view. The Phragmites wear their lifeless tan winter garb.

Duck-hunting season has begun, and if you listen closely in the morning, you can hear the sounds of shotguns from the Saw Mill Creek Wildlife Management Area echoing off distant landfills. The ducks in the region's tidal impoundments fly a few hundred yards away whenever a human approaches – with the Common Mergansers taking to the air at first sight of a human.

Just before Veterans Day, Mill Creek Marsh in Secaucus yields a host of winter sparrows – White-crowns, White-throats, American Tree Sparrows, Fox Sparrows, and Dark-eyed Juncos – joining a dozen late-departing Greater Yellowlegs, assorted duckage, a Ruby-crowned Kinglet, and the year-round feathered residents. The Spartina now sports its fall colors.

Bald Eagle sightings are a daily occurrence.

By month's end, the signs of winter are as visible as your breath in the morning, and looking through binoculars without gloves is often a challenge. The region's landscapes and marshscapes again are painted in tans and browns and several shades of gray.

Mehrhof Pond in Little Ferry has become Duck Central, with hundreds and hundreds of rafting Ruddy Ducks, plus an assortment of Ring-necks, Lesser Scaup, Northern Shovelers, and Hooded Mergansers. They will bide their time until the first big cold snap freezes the 20-foot-deep former clay pit and sends the ducks to the nearby open water of the Hackensack River.

Cedar Waxwings feast on ripe crabapples in DeKorte Park's Lyndhurst Nature Reserve in late November. *Photo by Dennis Cheeseman.*

December

The Meadowlands in early winter is a no man's land. Looking west across the river from Laurel Hill County Park, one can see wintering Great Blue Herons and Black-crowned Night Herons hunkered against the Phragmites that protect the birds from the hard west wind as they bask in morning sunlight.

An early December storm blankets the region in white, with the brutal winter winds creating snow drifts several feet deep. Throughout the month, the Northern Harriers, Long-eared Owls, and other raptors from Canada slowly trickle in and starting working the landfills and other prime hunting grounds.

The ducks are amassing in the tidal impoundments now, with the numbers of Canvasbacks increasing by the day. By mid-month, 14 species of ducks have made the Meadowlands their winter home.

A week later marks the start of the yearly Christmas Bird Count nationwide. The Meadowlands participants typically hold their local census on a Sunday, when MetLife Stadium is jammed with screaming (and freezing) fans. Some of the prime marshes for birding are located within earshot of the stadium, and the contrasts are striking – more than 80,000 people crammed into a small space, while thousands of acres of nearby wetlands are devoid of humans, save for a few bundled-up birders, trying to focus their binoculars despite numb fingers.

Toward year's end, the natural world has slowed. Life may not have gone into hibernation, but there's a slower metabolism to the world beyond our windows. To walk in the Meadowlands this time of year is a bit desolate, with strong winds making the dipping temperatures feel even colder, and a fleeting sun to the south taking any warmth with it.

Year's end is a time to be alone with one's thoughts, with fewer of nature's distractions, and fewer hours of daylight. A walk in nature in the late twilight helps us escape our occupations and preoccupations, our appointments and bills, our errands and to-do lists. It is time to think back to the Meadowlands year now behind us, and to look forward to the year that lies just ahead.

As for a New Year's Resolution, may it be this: May we never lose our sense of wonder at the nature that awaits us in the Meadowlands.

Common Mergansers start returning to the region in late November. *Photo by Ron Shields.*

A lone Great Blue Heron looks for some open water and a place to fish. *Photo by Marco Van Brabant.*

By December, birding can become an endurance sport. *Photo by Marco Van Brabant.*

Photo by Marco Van Brabant.

Acknowledgments

Text: The author would like to express his gratitude to New Jersey Meadowlands Commission Executive Director Marcia Karrow, Christine Sanz, Tom Marturano, Dr. Francisco Artigas, Dr. Ross Feltes, Mike Newhouse, Adam Levy, Brett Bragin, Gabrielle Bennett-Meany, Melissa Nichols, Brian Aberback, Donna Bocchino, Tammy Marshall, and the rest of the helpful NJMC staff; Michele Daly of the Meadowlands Environment Center, Don Smith, Joan Hansen, Erik Kiviat, Secaucus Historian Dan McDonough, Bob Ceberio, Lilo Stainton, Bruce and Karen Riede, Stephen Quinn, Ron Coldon, Gys Kooy, Kurt Muenz, the Meadowlands Museum, Cyndi Steiner, Jack Troy, Jeffrey Glassberg, Allen Barlow, Robert Kent, The Regional Plan Association, Kevin Karlson, and Russell Juelg and the New Jersey Conservation Foundation.

Photography and Art: The author expresses his gratitude to Ron Shields, Marco Van Brabant, Herb Houghton, Sanford M. Sorkin, Secaucus Historian Dan McDonough, Dennis Cheeseman, Regina Geoghan, Mike Girone, Michael C. Malzone, Sal Kojak and Brian Kennedy of the Meadowlands Environmental Research Institute, the Smithsonian Institution, the Newark Art Museum, New York Historical Society, Montclair Art Museum, Thomas F. Yezerski and Farrar, Straus and Giroux, Frank W. Zabransky and the Little Ferry Historical Society, The Associated Press, Kenneth Matteo of Central Jersey Pheasants Forever, Stephen D. McNamara of the Dawson Corporation, Harvey Sullivan IV, MacKenzie Hall of CWF, John R. Quinn, Jack R. Troy, Brent Thorne, Brett Bragin, Dennis Cheeseman, Jeff Nicol, Greg Miller, Joe McKay, EriK Kiviat, and the Meadowlands Museum, with a special thanks to LightHawk for its help in the aerial photography.

The author also expresses his gratitude to Mimi Sabatino of the NJMC for all of her expert advice and her help preparing the photography and art for publication.

Jim Wright, Ron Shields, and Marco Van Brabant produced the principal photography for this book. *Photo by Michael C. Malzone.*

Pilot Mike McNamara and LightHawk, a volunteer-based environmental aviation organization, assisted with aerial photography and helped the author get a better perspective of the district – and some great bird's-eye views.

Bibliography

BOOKS

Barlow, Allen E., David Michael Golden, and Jim Bangma. *Field Guide to Dragonflies and Damselflies of New Jersey*. Trenton, New Jersey: New Jersey Dept. of Environmental Protection, Division of Fish and Wildlife, 2009.

Brennessel, Barbara. *Diamonds in the Marsh,* Lebanon. New Hampshire: University Press of New England, 2006.

Cech, Rick, and Guy Tudor. *Butterflies of the East Coast*. Princeton, New Jersey: Princeton University Press, 2005.

Cawley, James and Margaret. *Exploring the Little Rivers of New Jersey*. New Brunswick: New Jersey: Rutgers University Press, 1993.

Facciolla, Nicholas W. *Minerals of Laurel Hill*. Secaucus, New Jersey: *s.n.*, 1981.

Glassberg, Jeffrey. *Butterflies Through Binoculars (Boston – New York – Washington Region)*. New York, New York: Oxford University Press, 1993.

Gochfeld, Michael, and Joanna Burger. *Butterflies of New Jersey*. New Brunswick, New Jersey: Rutgers University Press, 1997.

Karlson, Kevin. *Birds of Cape May*. Atglen, Pennsylvania: Schiffer Books, 2010.

Kiviat, Erik. *Monitoring Biological Diversity in the Hackensack Meadowlands*. Lyndhurst, New Jersey: Meadowlands Environmental Research Institute, 2007.

Leiby, Adrian C. *The Hackensack Water Company 1869-1969*. River Edge, New Jersey: Bergen County Historical Society.

Quinn, John R. *Fields of Sun and Grass*. New Brunswick, New Jersey: Rutgers University Press, 1997.

Rezendes, Paul, and Paulette Roy. *Wetlands*. San Francisco, California: The Sierra Club, 1996.

Royte, Elizabeth. *Garbage Land*. Boston, Massachusetts: Little, Brown, and Company, 2005.

Sullivan, John Langdon. *Report and explanation of a survey for a rail-road from Paterson to New York*. Society for Establishing Useful Manufactures, 1829.

Sullivan, Robert. *The Meadowlands*. New York, New York: Scribner, 1997.

Torrey, John. *A Catalogue of Plants Growing Spontaneously within Thirty Miles of the City of New York*. Albany, New York: Lyceum of Natural History of New York, 1819.

Vileisis, Ann. *Discovering the Unknown Landscape*. Washington, D.C.: Island Press.

Weis, Judith S., and Carol A. Butler. *Salt Marshes*. New Brunswick, New Jersey: Rutgers University Press, 2009.

Wright, Kevin. *The Hackensack Meadowlands*. Lyndhurst, New Jersey: The Hackensack Meadowlands Environment Center, 1988.

PERIODICIALS

Cunningham, John T. "Monumental Meadows." Newark: *Newark Sunday News*, series, February 8 – April, 5, 1959.

Ehrenfeld, Joan G. "Evaluating wetlands within an urban context." *Ecological Engineering 15*, 2000.

Gansberg, Martin. "Federal Agency Praises Cleanup of Once 'Stinking' Meadowlands." New York, New York: *New York Times*, May 8, 1978.

Kiviat, Erik, and Kristi MacDonald. "Biodiversity Patterns and Conservation in the Hackensack Meadowlands, New Jersey." *Urban Habitats*, December 2004.

Unknown author. "The Improved Highway Across the Hackensack Meadows." Letter to the Editor, *The Horseless Age*, December 9, 1903.

Unknown author. "Hackensack Meadows a Hiding Place for Fugitives." New York, New York: *New York Times*, August 28, 1910.

Unknown author. "A Magic City from a Swamp." *Popular Science Monthly,* October 1928.

Unknown author. "Series of Multivehicle Collisions and Fires under Limited Visibility Conditions, New Jersey Turnpike, Gate 15 and U.S. Route 46, October 23 and 24, 1973," *Highway Accident Report*. Washington, D.C.: National Transportation Safety Board, Adopted March 5, 1975.

O'Neill, James M., "Meadowlands bee variety has experts buzzing." Woodland Park, New Jersey: *The Record,* July 9, 2010.

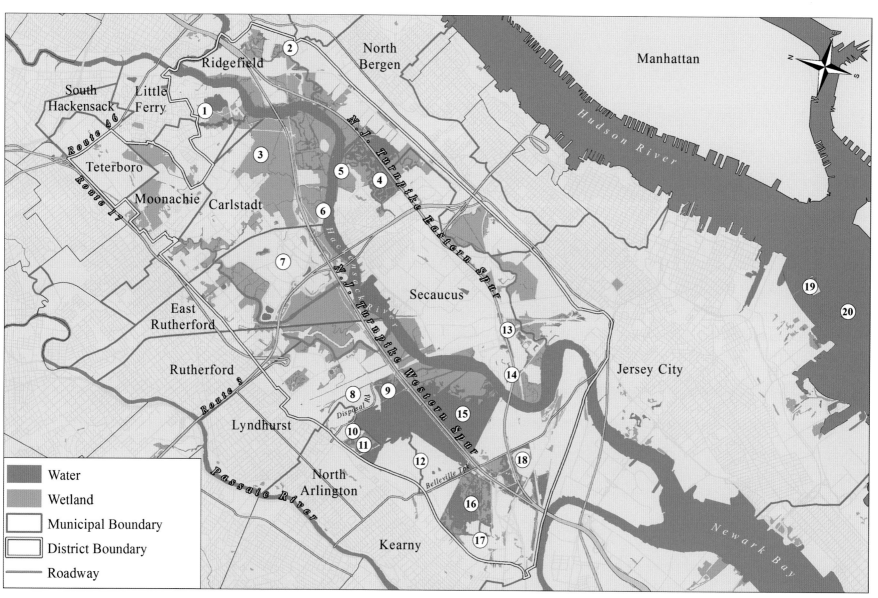

Water
Wetland
Municipal Boundary
District Boundary
Roadway

1 Losen Slote Creek Park, Little Ferry
2 Skeetkill Creek Marsh Park, Ridgefield
3 The Richard P. Kane Natural Area, Carlstadt
4 Mill Creek Marsh, Secaucus

5 Mill Creek Point, Secaucus High School Marsh and Boardwalk, Secaucus
6 River Barge Park, Carlstadt
7 MetLife Stadium and Meadowlands Sports Complex, East Rutherford

8 Kingsland Landfill, Lyndhurst
9 Richard W. DeKorte Park, Lyndhurst
10 Erie Landfill, North Arlington
11 Harrier Meadow, North Arlington
12 1-E Landfill, North Arlington, Kearny

13 Secaucus Junction Rail Station, Secaucus
14 Laurel Hill County Park, Secaucus
15 Saw Mill Creek Wildlife Management Area, Lyndhurst, North Arlington, Kearny

16 Kearny Fresh-water Marsh, Kearny
17 Keegan Landfill, Kearny
18 1-A Landfill, Kearny
19 Ellis Island
20 Statue of Liberty

Index

A Question Mark perches near Jill's Butterfly Garden in DeKorte Park.

Midges by the thousand have replaced mosquitoes as the Meadowlands most annoying insect. The good news they don't bite. *Photo by Marco Van Brabant.*

DeKorte Park's Lyndhurst Nature Reserve after a snowstorm. *Photo by Marco Van Brabant.*

A Great Blue Heron gets a great view of the setting sun.
Photo by Marco Van Brabant.

A Salsify's seed head resembles that of a giant Dandelion. *Photo by Marco Van Brabant.*